Screen Society

Ellis Cashmore · Jamie Cleland
Kevin Dixon

Screen Society

Ellis Cashmore
School of Languages and Social Sciences
Aston University
Birmingham, UK

Jamie Cleland
UniSA Business School
University of South Australia
Adelaide, Australia

Kevin Dixon
School of Social Sciences, Humanities and
Law
Teesside University
Middlesbrough, UK

ISBN 978-3-319-68163-4   ISBN 978-3-319-68164-1   (eBook)
https://doi.org/10.1007/978-3-319-68164-1

Library of Congress Control Number: 2018937875

© The Editor(s) (if applicable) and The Author(s) 2018
This work is subject to copyright. All rights are solely and exclusively licensed by the Publisher, whether the whole or part of the material is concerned, specifically the rights of translation, reprinting, reuse of illustrations, recitation, broadcasting, reproduction on microfilms or in any other physical way, and transmission or information storage and retrieval, electronic adaptation, computer software, or by similar or dissimilar methodology now known or hereafter developed.
The use of general descriptive names, registered names, trademarks, service marks, etc. in this publication does not imply, even in the absence of a specific statement, that such names are exempt from the relevant protective laws and regulations and therefore free for general use.
The publisher, the authors and the editors are safe to assume that the advice and information in this book are believed to be true and accurate at the date of publication. Neither the publisher nor the authors or the editors give a warranty, express or implied, with respect to the material contained herein or for any errors or omissions that may have been made. The publisher remains neutral with regard to jurisdictional claims in published maps and institutional affiliations.

Cover credit: Joel Sorrel/Getty Images

Printed on acid-free paper

This Palgrave Macmillan imprint is published by the registered company Springer International Publishing AG part of Springer Nature
The registered company address is: Gewerbestrasse 11, 6330 Cham, Switzerland

Contents

Screen Society: Timeline

1696
Demonstration in England of a device that will become known as a magic lantern, a primitive form of image projector used for showing images on a screen.

1839
Louis-Jacques-Mandé Daguerre presents the first practical photographic process to the French Academy of Sciences, establishing the technology that will yield practical photography. In the same period, magic lantern shows are commonplace in Europe and America.

1850s
Theatrical entertainment grows in popularity, principally through a music hall, which offers a mix of songs, comedy, conjuring and speciality acts. Audiences sit in spacious theatres with seats and a stage on the understanding that they respect others' comfort. GAMECHANGER

1895
In Paris, Auguste and Louis Lumière exhibit their kinetoscope, which lets viewers view moving images through a peephole. This starts a period

of rapid experimentation, leading to the development of the cinematograph, which allows simultaneous viewing by multiple people.

1910s

The popularity of the cinematograph encourages the production of silent films to entertain paying audiences. By the end of the decade, there will be a fledgling film industry. GAMECHANGER

1921

The Kid, a silent black and white film featuring Charlie Chaplin is released.

1922

BBC Radio launches. Radio will become the dominant broadcast medium of the period.

1927

The Jazz Singer is released. This is the most successful film with sound to date.

1930

Film is established as a mainstream entertainment, supplanting theatre.

1936

BBC launches Britain's first regular tv service at 3 p.m. on November 2, but only for two hours per day and to London audiences only.

1937

Walt Disney Studios (later to become Disney) release their first feature film, *Snow White and the Seven Dwarfs*; an animated production that offers a new type of family entertainment.

1939

Gone with the Wind, the 221-minute epic film based on Margaret Mitchell's novel, sets new standards in cinema and will go on to become one of the highest grossing films (when adjusted for price inflation).

1941

Konrad Zuse creates the first functional electro-mechanical binary programmable computer, a forerunner of today's devices.

1949
BBC televises beyond London.

1950s
Sales of domestic television rise dramatically. GAMECHANGER

1952
Cinema responds to the challenge to its popularity with colour stereo-scopic 3D and CinemaScope.

Bell Labs, the American research and scientific development organization, invent the modem, a device that converts digital signals back to electrical signals and back again, enabling communication between computers.

1951
I Love Lucy, which started as a radio comedy, is turned into a television sitcom and airs on American network CBS. The series will run till 1957 and will be screened internationally. Such is the popularity of the show that it will greatly assist the rise of television as a genuine challenger to cinema.

1960
The first universal standard for computers ASCII (American Standard Code for Information Exchange) is developed and permits machines from different manufacturers to exchange data.

1963
Abraham Zapruder films the assassination of US President John F. Kennedy on his Model 414 PD Bell & Howell Zoomatic Director Series camera. The 26.6 seconds of footage is the only film of the shooting and will assume iconic status in the years that follow. Zapruder's camera is one of the home movie cameras that are popular in the early 1960s. The films are developed and shown on portable screens years before video (see below, 1978).

1965
Polaroid's Swinger camera allows users to see photographs without sending them off to the chemist for development.

1967

Decca introduce colour television with its model CTV 25.

1969

An estimated 600 million people watch the Apollo 11 landing on the moon live on television, a world record until 1981, when 750 million people watch the wedding of the Prince of Wales and Lady Diana Spencer.

A computer-to-computer link is established on the Advanced Research Projects Agency Network, or ARPANET, which is funded by the US government. The first transmission is sent by a UCLA student programmer to Stanford Research Institute's host computer. Fifteen years later, the first email will be sent. GAMECHANGER

1970

The first IMAX screen is unveiled in Osaka, Japan.

1971

Ray Tomlinson implements the first email programme on the ARPANET system, a precursor to the internet.

1973

Flexible removable, magnetic disks, popularly known as floppy disks enable users to save and transport documents, though a whole disk can't store a whole song.

CEEFAX, a text information system is integrated into television sets.

1975

Sony introduces the Betamax machine: this records and plays back television programmes on plastic cassettes (about as big as hardback books), enabling viewers to tape and watch programmes at their own convenience.

1976

JVC release the first video home system, or VHS, machines in Japan (then in the UK and USA in early 1977). This rivals Sony's Betamax (see above, 1975).

1980

The Sony Walkman cassette tape player is launched. Sales will reach 200 million worldwide.

1981
The Acorn BBC Micro Computer is used in schools.

1982
Kilnam Chon, a professor at Keio University in Japan, develops the first internet connection in Asia.

1984
Apple Inc. introduces the computer mouse. GAMECHANGER
  The first email arrives in Germany from the US on August 3.
  Van Jacobson solves internet congestion by developing algorithms for the Transmission Control Protocol. They are still used in over 90% of internet hosts today.
  Brewster Kahle invents the first internet publishing system. It is the precursor to today's search engines.

1986
Digital cameras are available at retailers, though most models only hold 6 pictures.

1987
The film *Wall Street* features Michael Douglas as "Gordon Gekko," who uses a brick-like mobile phone, the Motorola DynaTac 8000X. It weighs two pounds, or 902 grams, and is over a foot long (32 cm). The phone had been on the market since 1983, when it was priced about $4000.
The Finnish company Nokia introduces a mobile phone called the Mobira Cityman 132. It weighs 1.7 lbs, or 760 grams, and is 7 × 3 × 1.7 inches, or 18 × 8 × 4 mm.

1989
World Wide Web (www) begins as the European Organization for Nuclear Research, better known as CERN, initiates a project called ENQUIRE. GAMECHANGER
American Online (AOL) launches its Instant Messenger chat service and uses the now-famous greeting "You've got mail."

1990
Tim Berners-Lee develops a computer programme with a graphical user interface for displaying Hypertext Markup Language, or HTML files,

which is used to navigate the World Wide Web. This becomes known as a browser. The term "surfing the net" becomes popular over subsequent years.

Nintendo Game Boy, the handheld games console, is introduced, giving rise to what becomes known as the "Tetris Effect," in which gamers have hallucinations of slotting bricks after hours of playing. It takes its name from the video game *Tetris* and is probably the first disorder attributed to spending time in front of a gaming screen.

1991

Phil Zimmerman creates email encryption software package that is published for free. It becomes one of the most widely used packages.

The World Wide Web is made available to the public for the first time on the internet.

The Lithium-ion battery makes lighter, rechargeable gadgets possible.

1992

The laptop makes its first appearance, liberating office workers from the shackles of the desktop computer. IBM's 300 Thinkpad weighs 6 kg, or 13.3 lbs.

StarWorks offers a commercial on-demand video service originally known as "store-and-forward" video, but which is later called streaming. Six years later, Netflix will appear on the market (see below, 1998 and 2007).

The first ever text message is sent when a British engineer uses his computer to send "Merry Christmas" to an Orbitel TPU 901, a mobile phone weighing 2100 grams, or about 4.6 lbs. This is made possible by SMS, or short messaging service. By 2017, 200,000 text messages are sent every every second.

1993

CERN makes its World Wide Web technology available in the public domain.

The computer system, Windows 3.1 means that users could click on pictures and icons rather than typing demands on a keyboard.

1994

Stanford University graduate students Jerry Yang and David Filo create Yahoo!

1995
Amazon.com, advertised as "The World's Biggest Bookstore" opens.

The world's first online dating agency, Match.com, begins.

Playstation1 sells 100 million consoles worldwide. Its popularity lead many to assume it has addictive properties.

1997
Nokia introduces a light, portable mobile phone, the 6110.

Google.com registers it name as a domain (i.e. a distinct subset of the internet under the control of a particular organization or individual).

1998
The first blog (i.e. weblog) appears spurs on by the advent of web publishing tools that have become available to non-technical users.

Netflix offers a new mail order service: sending DVDs to homes.

DVD Players allow customers a high quality cinematic experience in their own homes. Higher quality than video's anyway!

1999
Nokia continues to lead the mobile phone market, introducing its 3210 model, which can fit in the palm of a hand. 160 million 3210s are sold and the model remains in circulation into the twenty-first century.

Craig Newmark founded Craigslist which changes the way people use classifieds, transforming it into a largely internet based industry.

2000
AOL acquires Time Warner for $165 billion, creating the world's biggest media organization (the company will split into two after ten years when the original companies resume trading independently).

Aaron Swartz co-creates RSS, a program that collects news from various websites and puts it in one place for users.

2001
Jimmy Wales starts the collaboratively written online encyclopaedia Wikipedia. Contributors become known as Wikipedians; within a year, there are 20,000 entries.

iPods are introduced to the market, making the Sony Walkman (see above, 1980) seem primitive by comparison. Music can now be stored digitally on a portable device.

2002

Microsoft launches its Xbox, an online multiplayer gaming service.

Mobile emails: BlackBerrys allow users to check emails on the go.

Camera phones become widely available making selfies inevitable. In 2016, 24 billion *selfies will be uploaded to Google.*

2003

Apple's iTunes opens for business, offering 200,000 tracks for a price.

Skype launches.

2004–2005

Web 2.0 enables interactivity. GAMECHANGER

Facebook is founded. By 2014 the social networking site will have 1.2 billion users. GAMECHANGER

The SatNav is brought to the market in 2004, revolutionizing the road trip by devolving responsibility for navigation to the GPS device.

2005

YouTube launches using user-generated content—the hallmark of Web 2.0. Its very first video is still available for viewing here: https://www.youtube.com/watch?v=jNQXAC9IVRw.

Broadband surpasses dial-in connections for the first time.

2006

Twitter is launched.

High definition television arrives, screens go flat.

Spotify gives users access to millions of tracks for free.

2007

Apple introduces the iPhone (with OSX), a device that allows watching films, listening to music and browsing the internet as well has having a 2 mega-pixel camera. Its effects are manifold: by 2017 1.2 trillion digital photographs per year will be taken, 85% on phones. This is the first type of smartphone. GAMECHANGER

With its mail order DVD business faltering due to competition from, among others, Apple, Netflix launches a revolutionary concept: delivering movies directly to consumers' computers.

BBC iPlayer let viewers watch programmes whenever they wish, presuming they have broadband.

2008
Apple launches its first App store with 500 applications.

Music Streaming grows in popularity. Revenues will rise from zero in 2007 to £125 million in 2014.

2009
Uber starts, using an on-demand taxis app.

Advertising online surpasses traditional forms.

The fitness tracker Fitbit becomes available.

Kindles are available to buy, enabling users to download thousands of books on one portable device made by Amazon.

2010
Instagram, the photo-sharing application and service that allows users to share pictures and videos, is launched. By 2017, it will have 700 million users.

China dominates internet usage with over 450 million internet users.

The first iPad tablet is designed, developed and marketed by Apple. It is 13 mm, about a half-inch thick, and will help Apple become, by 2017, the world's most valuable company with a value of $170bn.

2011
Live streaming of Prince William and Kate Middleton's wedding is the most-watched event on the internet to date.

2013
Twitter launches a music app.

2014
City authorities in Chongqing, southwest China introduce a dedicated 30 meter (99 feet) walking lane for pedestrians who habitually stare at their phones or other screen devices. The newly coined condition "distracted walking" is recognised as an omnipresent feature of Screen Society (see below, 2017).

2015
Apple's Smartwatch, which is oriented to health capabilities and integrates with other Apple products is launched.

2016
The Smart Hub arrives: the Amazon Echo answers questions, plays music, reports the news and is thought to be the forerunner to artificial intelligence in the home.

2017
Pedestrians looking at the screens of phones or tablets while crossing the street can be penalized after Honolulu becomes the first major city to pass legislation aimed at reducing injuries and deaths from "distracted walking." Texting while driving is already illegal in many parts of the world.

Chewing gum sales fall 15% since 2007, the year in which the iPhone is launched. The speculative, though plausible explanation being that supermarket checkout queues have chewing gum displays to tempt consumers into buying as a way of staving off boredom. With smartphones, there is less boredom and hence less impulse to purchase the gum!

1

Introduction

A Screen-Less World

Suppose we had no screens, those flat panels on which images and data are displayed. Electronic devices, such as televisions, computers and smartphones have them. And when we go to the cinema, we see gigantic screens, the biggest IMAX being over 35 metres (100 foot) tall. Every city has advertising hoardings, or billboards, which used to be printed on card, but are now more likely to be digitally projected on huge screens, the online betting company Betfair boasting one in Vienna the size of 50 football fields.

Wherever you are, you can probably raise your head, look around and see some kind of screen in your immediate environment; that's in addition to the one you're carrying. Screens are so ubiquitous and inescapable that we barely notice them. Try to find public space, whether in a bar, restaurant, department store, or in the street where there is no screen pleading for your attention with moving text and images.

History would be different in the parallel screen-free universe: one in which no one came up with the idea of projecting images onto a blank surface in the early sixteenth century and no one saw the potential

© The Author(s) 2018
E. Cashmore et al., *Screen Society*, https://doi.org/10.1007/978-3-319-68164-1_1

in turning this into a way of distracting us in an agreeable way in the seventeenth. We wouldn't have been entertained by the magic lantern, as it was called, and we wouldn't have been captivated by moving images called motion pictures in the early twentieth century. And we wouldn't have had our culture transformed in the 1950s by arguably the most influential invention in history: television.

Television changed culture and, by implication the people who create culture—we humans. The idea of not having to travel to and gather at public places to be entertained by sound and image had far-reaching effects on practically every aspect of our lives. In its day, early tv sets were like portable Aladdin's caves: instead of going somewhere to find a place filled with an exotic miscellany of strange and precious items, we could have a cave of our own; even better, we could take it with us wherever we went.

From the concept of a screen that's our own possession and which we can use whenever and wherever we choose, we've fashioned any number of portable devices. Personal computers arrived in force in the 1990s. Then in 1997, Nokia introduced its 6110 model phone, which was light enough to carry around. And the merger of phones and computers brought us smartphones, Apple's first iPhone arriving in 2007.

How could we cope without them? What would we do first thing in the morning if not check our email inbox? How would we communicate without sliding and tapping our fingers? From where would we get our information, including world news, if the particulars of events weren't right in front of us? How could we organize our days and nights without a constant flow of instruction about who's going to where and when? Perhaps most fundamentally, how could we sustain social life without them? We've created and maintain a culture in which we live through and depend on media. And we access that media through our screens.

Think about how you get your knowledge: your facts, information, intelligence and understanding of a subject. Obviously, we talk to each other face-to-face, though scholars and politicians often complain that we don't do enough of this. They probably don't grasp that communicating via the phone or tablet is as rewarding and meaningful as standing next to someone and exchanging thoughts. Often it is more enjoyable. This is one of those basic points that's frequently missed by self-appointed authorities who pontificate on the uses and abuses of digital media. People use their devices for communicating because they enjoy it. Simple but true: users derive pleasure from using their devices. If they didn't, they probably opt to communicate via different methods, or communicate less.

The pleasure people take from their computers, phones and tablets is, like many other types of pleasure, not necessarily intelligible to those who are not habitual users. In this sense, it's like music. Some people will listen to hip-hop and scratch their heads in wonder why it's one of the world's most popular genres. The minimalist music of Steve Reich or Philip Glass some listeners will find tedious, while others will rhapsodize over the different musical languages and decree that outsiders just don't get it. As we'll argue later in this book, many critics of social media are not just like but actually *are* outsiders, who are trying to fathom out a new language word-by-word, but without any understanding of its grammar.

In a screen-less world, it's difficult to fathom how we'd learn about practically anything. And by learn, we don't mean learn in a narrow academic sense, but in the broader sense of becoming aware by receiving and transmitting information. Everything we know and much of what we do is mediated. It's connected through other people or things. It involves an intermediate agency. How could it not? We couldn't possibly experience first-hand everything we know about the world. There never was a time when people did that. There's always been a category of people, like messengers, town criers, or things, like newsheets, books, and, before them, scrolls or even wall drawings such as hieroglyphs in the ancient world. Most knowledge is mediated in some way.

Yet there is something different about today. There's never been a time in history in which we spend so much time engaged with the media and rely on it to such an extent, not only for our knowledge but for our friendships. Print media has been with us theoretically since the mid-fifteenth century, when the German printer Johannes Gutenberg developed movable type on the machine known as the press. The term became shorthand for print media. Four hundred years later, the world relied on the press for its information about almost everything.

Print media made demands on us: the ability to read being the principal one. It became, with printed books, one of the catalysts of literacy. To understand the content of newspapers, gazettes, magazine, newsheets and the several other forms of press the consumer needed a working knowledge of the written word. Radio made no comparable demands on its consumers: they could just listen to spoken words. From the early twentieth century, sound messages carried information to us through electromagnetic waves and the transmission became known as broadcasting.

Like television, which followed in mid-century, it tended to tax consumers less: broadcasting information required consumers to listen or look and to think, though not necessarily concentrate in the way they would when reading. It took until the late twentieth century before university scholars argued persuasively that listening to radio or looking at tv required cognitive action or interpretive skill comparable with reading a written text. In fact, the output of radio and tv was actually called text and the process of making sense of it was called—in the manner of rendering the written material comprehensible—reading.

Amusing Ourselves

Think about the very concept of watching a screen. Audiences in the eighteenth century, or possibly before, would have gasped at the images they saw projected onto blank screens by the invention known as *laterna magica* and we will trace this history more thoroughly in Chapter 2. They would have probably suspected some kind of magic or a diabolical deception to induce their attention. Three hundred or so years later and there are still people who insist our fixation on screens is the devil's work.

Twentieth century audiences, as we'll also discover in later chapters, were used to big screens. The first known cinema was built in 1894 and movie theatres sprung up across Europe and North America in the following decades. Sitting in a crowded auditorium was not uncomfortable for audiences brought up on theatre, music hall and, in the USA, rag-time. Yet the gigantic stationary, flat, two-dimensional screen was a big change from a stage populated by live performers.

Readers of this book will not have known a time when television did not have a prominent presence. Anyone born before 1940 may have a recollection of the age before television, though the majority of their lives will have been lived in an environment that will have been massively affected by television. Practically every habit was, in some way, influenced by our captivation with tv.

In a way, film prepared audiences for television. Goggling at a 12-inch (30 cm) diameter tv screen (that was the size of the early models) was actually not so different from staring at a cinema screen. The big difference was that audiences were obliged to keep quiet while a film was playing and couldn't dictate when to switch on or off (though they could always walk out, of course). But televisions were portable: tv sets were either bought or rented, so they were effectively our possessions; our own private screens. Television ownership soared in the 1950s. By the early 1960s, no home was complete without at least one set. For a while, it seemed it would wreck the film industry. But what would it do to us?

The scares about television were many: watching tv would shorten our attention spans, delimit our social abilities, break down families, affect our propensities, particularly to violence, and so on. There were dozens of possible harmful effects. But no one seemed interested as programmes proliferated and sales of domestic sets climbed.

There was what seemed a wilful disregard of the informed opinion of the time (we should bear this in mind when we think about today's habits). Television was seen as one of the most menacing developments around. It cultivates abnormal relationships, pins us in our homes and nurtures passivity, said critics. One of the most brilliant books on television, written in 1985 by the American media critic Neil Postman, had the title *Amusing Ourselves to Death*. The idea being that we were feasting on too much entertainment. And, the book's justifiable assumption was that television is good for only one thing: entertaining us. This supposition is worth unpicking.

Every time we turn to our screens we expect to be entertained. If not we're disappointed. Obviously entertainment has to be entertaining, but nowadays, so does politics, crime, health reports, and so on. If they don't entertain us, we dump them. By entertain, we mean engage us in a way we find agreeable, even better enjoyable. Notice we don't include words like superficial, shallow or trivial in our definition of entertain. Some forms of entertainment might be all of these, but other forms require serious thought and deep consideration. We can learn at the same time we're being entertained.

Entertainment might be regarded as a mode for everything: a way or manner in which politics, crime, health, education, even religion are expressed and experienced. If anything is going to get our attention for any length of time it had better be presented in a style that engages us agreeably. Television started with different ambitions. In the US, it was intended to be an extension of radio, which itself was an advertising medium; the programmes were merely to catch and keep listeners rapt.

In Britain and other European nations, tv was launched with loftier ambitions. BBC television was, like its American counterpart, a descendant of radio; but radio in Britain carried no advertising and was never envisaged as having commercial value. Rather it was meant to contain quality arts programmes, major documentaries about history and culture, and large-scale live coverage of major national events and anniversaries. A theatre of the airwaves, as it was known. Television in both the US and the UK and everywhere else in the world, succeeded because it was supple and flexible in its design and adapted effectively to suit changing environments. Actually, tv didn't just adapt to environments: it became a catalyst in instigating changes. It was the captivating medium not only of the twentieth century but of all time.

There had never been a phenomenon like television for inciting peoples' attention. In 1969, 530 million people, that's 14% of the population of the world at the time, watched the moon landing. Even this seems modest compared to the estimated 2.5 billion who watched in some part the funeral of Diana, Princess of Wales, in 1997. These were unique events, though some sports events, particularly football's World Cup Final and the Super Bowl regularly attract hundreds of millions. Television arrived after cinema, but was much more influential in inculcating audiences into the habit of staring at screens while they made sense of the unfolding narratives. Postman was a piercingly intelligent critic of television, but even he couldn't accept that the cognitive work required when watching tv was comparable with that needed to engage with other media.

Television had no competitors up till quite recently. The fortunes of the film industry fluctuated, though it survived tv's initial onslaught, then withstood pressure from home videos, DVDs and piracy. It remains afloat. For a period, it appeared television too would be under threat from social media sites like Facebook and YouTube. The latter in particular sent a frisson through the advertising industry when the numbers were revealed. A YouTube star like Zoella could remain anonymous in traditional media but command 6.5 million subscribers on her YouTube channel. Companies such as Pepsi started to advertise more on digital platforms than they did on conventional media. Since 2010, the amount of television watched by those aged 16–34 has fallen steadily. But, far from going under, traditional tv has prospered from the internet, sharing platforms with streaming providers and subscription broadcasters, so that its content can be consumed on a variety of portable devices, not just the home appliance. It's probably inaccurate even to call it traditional television nowadays: there are so many ways to view television that there is little traditional about it. In fact, television remains an integral part of the *Screen Society*.

Some say it was the defining invention of the twentieth century: not only did it change our social habits and our cognitive abilities, but it made the world smaller. News of events in any part of the world could circulate, at first in days, later in hours, and eventually in minutes, thanks to rapidly changing technology. Television also induced a reliance that we may not have shrugged. It became the main source for news and current affairs as well as entertainment. Cinema, theatre, nightclubs, bars: none of them had magic strong enough to rival television's.

Its relevance to the current century is uncertain. But television's legacy will be felt for the next several decades. It was the device that habituated us to screens. We became habituated to them very easily, it seemed. Despite the warnings, we accepted television as we might welcome a new friend who brought with her or him an endless trove of delightfully amusing treasures. Almost paradoxically, that meant staring for hours and hours at a limited surface. We still do this. The big difference is that we can now carry the screens around with us.

Were we speculating on this development twenty years ago, we might have argued persuasively that consumers will use the portable screens as and when they needed to watch a show, an item of news or some other presentation they wished to enjoy. The implication would have been that we will not binge, that is indulge excessively in the activity. Who could imagine a world in which people are constantly holding their screens in front of them, gazing perpetually while they attempt other, sometimes tricky endeavours, like climbing stairs, walking through crowded shopping areas or trying to concentrate on a lecture? Twenty years ago, remember. Today, this is a reasonable description of how we live.

Bad for Society?

Does this mean digital media is bad for society? This is a crass and value-loaded question, but it's intended to be: the answers are usually just as crass and value-loaded. Consider the following selection of headlines (mostly from traditional media) that issue warnings about the consequences of our current engagement with social media.

UNDER-5s GLUED TO SCREENS 4 HOURS A DAY
(*Daily Mail*, 15 November, 2016)
ELECTROSHOCK THERAPHY FOR INTERNET ADDICTS
(*New York Times*, 13 January, 2017)
FACEBOOK, TWITTER AND GOOGLE HAVE BECOME A 'RECRUITING PLATFORM FOR TERRORISM'
(*Telegraph*, 25 August, 2016)
HOW TO STOP CHECKING YOUR SMARTPHONE IN THE MIDDLE OF THE NIGHT
(*Telegraph*, 26 September, 2016)
WE SPEND 1.3 YEARS OF OUR LIVES DECIDING WHAT TO WATCH ON TV

(*ShortList*, 3 November, 2016)
FRAUDSTER ADDICTED TO TV SHOPPING STOLE £370,000
  FROM HER EMPLOYER
(*Coventry Evening Telegraph*, 9 January, 2017)

We'll investigate all these arguments and the assumptions or research that informs them later. There is certainly a formidable body of opinion and research that personal and social life is suffering as we turn away from each other and towards our devices. Perhaps the most suggestive contribution of recent years is the report published in December 2017 in the *Journal of Health Psychology* by a team of researchers from the University of Oulu, Finland, and Nottingham Trent University, England. This report focused specifically on video games, a subject we cover in Chapter 9, and which the researchers argue, "have increasingly become an integral aspect of individuals' leisure activities and everyday life."

The researchers accept the *Diagnostic and Statistical Manual of Mental Disorders*, 5th edition, which includes "internet gaming behaviour," or IGB, and wish to extend this with their own "problematic gaming behaviour," or PGB, which is "a behavioural pattern encompassing persistent and recurrent engagement with both online and offline games, leading to significant impairment or distress," as they put it on page 2 of their article. (We'll detail all the research and other kinds of publications we quote or cite at the end of each chapter.)

The research team led by Niko Männikkö found PGB has several "adverse health-related outcomes." It contributes to depression, anxiety, low self-esteem and many physical ailments including cardiovascular stress, wrist pain, issues with sleep and the nervous system. Gamers are also at risk of mental side effects, ranging from obsessive-compulsive behaviour, a lack of concentration and self-control and impulsiveness. It's a prodigious list of infirmities, considering gaming is a leisure activity (and a professional sport, actually) that's meant to be enjoyed.

The report is statistically detailed and examines data from more than 130,000 gamers, aged between 12 and 88. The problem is: we don't hear from any of them: no space is allowed for their own accounts, apprehensions and experiences. The report is what's known as a meta-analysis, meaning it collects data from 50 other studies into video game addiction, all similarly uncritical of the orthodox view that there is something inherently problematic and addictive about gaming. It may not be a typical study, but it does reflect the prevailing convention in studies of this kind. Most are conducted by psychologists, neuroscientists and health researchers. Few seem interested in the social, cultural or historical contexts in which gaming takes place, or the perspectives of the people who engage in the activity themselves—the gamers. As such, they present the view of "experts," not users. We believe there is much to learn from the users—and much to criticize when we're offered the findings of "experts."

For example, smartphones have been singled out for promoting sleep deprivation and mental health disorders. There's no doubt we do stare at screens a lot: wherever you go next, look around you and notice how many others are staring at computers, phones or tablets; and how many others are talking on the phone; and how many others are listening to something through their ear pods. It might be easier to spot those who are not. But is this practice as threatening as the headlines and the studies indicate? After all, similar warnings were sounded but went unheeded when television started to invade our lives to the point where we became what Brits still call "telly addicts."

But overuse is a misleading term in this context: it means people use their smartphones too much. In the same way, many of us use the phrase "you know" too much in our conversations; or we use our cars too much for short journeys instead of walking. This doesn't mean we couldn't improve our conversational skills by using "you know" prudently, or shouldn't contribute towards saving the planet by sparing the environment CO_2 fumes. But, when we apply the phrase to smartphones, the questions arise: what constitutes overuse and who or what benefits from cutting back on using smartphones? This is where the debate slides between facts and values, science and morality. There are no absolutes in this debate. It seems that, as with television, a practice has become so universally popular in such a relatively short period of time that self-appointed experts have enthusiastically but perhaps mistakenly assumed there are problems. Something so popular and so clearly enjoyable and which confers so much pleasure to so many must have a downside.

Apples and Oranges

Barely a day goes by without a new piece of research either issuing cautions or giving assurances about our media habits, usually the former. For example, in his 2017 book *Irresistible: The Rise of Addictive Technology and the Business of Keeping Us Hooked* Adam Alter, a professor at New York University, warned connectivity threatens the health of not just our children, but everyone. He described a scenario that could have been taken straight from the tv series *The Wire*, which depicted how young children are given crack as a gift, just to get them into the habit. "As a kid I was terrified of drugs," Alter remembered. "I had a recurring nightmare that someone would force me to take heroin and that I'd become addicted."

Alter was careful to distinguish between an addiction, which he argued is an indulgence which brings pleasure, and a compulsion, which he contended is an indulgence which merely brings relief from restless anxiety (a distinction that we will question in Chapter 4, in which we will offer a different perspective to that of Alter).

But consider the results of earlier research from The Pew Research Center's Internet and American Life Project, where Keith Hampton and his colleagues painted a rather different and more complex role that digital media play in people's lives, emphasizing the positive impact of the widespread use of social networking for establishing new relationships based on, among other characteristics, trust, tolerance, support and political engagement. In this 2011 study, the media, far from being frightening and addictive, offered plenty of benefits. There was no evidence that users were any more likely than others to become inured in particular habits by cocooning themselves in social networks of like-minded, and perhaps similarly addicted people—as Alter and many others fear.

There's no formula for comparing studies such as these; in fact, some readers might think we are comparing apples and oranges. In other words, they're considering different aspects of media use. It's conceivable that someone could be hopelessly addicted to the net because it keeps bringing them new, rewarding relationships. Conceivable, perhaps;

but unlikely. Perhaps both pieces of research should be understood as revealing the *potential* of digital media: screens can, but will not necessarily prove destructive but, equally, they might, though not always, improve wellbeing, satisfaction and enjoyment.

Potentially is a good term to keep in mind when thinking about anything associated with digital media and its consequences: like any form of technology it has capacity to develop once in the hands of its users. Technology may change us. But, more importantly, we change technology. A smartphone is loaded with inherent and possibly unrealized capability. That's all. And, in a sense, it's been forever thus. Let's take another slight historical detour to explain our meaning.

Revolution

Even before television, there were probably plenty of opinion leaders who warned of the dangers of film, and, before them, doomsayers who prophesied the problems associated with entertainment generally; they probably thought too much frivolity was incompatible with an earnestly-led religious life. Certainly, the industrialism that transformed western society and, eventually, the world in the late eighteenth and early nineteenth centuries inspired many warnings. New technology and industrial development had convulsed life, drawing populations to cities in search of work. The factories in which people were obliged to work posed obvious dangers; machine accidents became commonplace. People were herded together in uniformly ordered streets and lived in close proximity to each other in a way that contrasted with the agricultural patterns of life. According to observers of the time, a new type of mentality emerging in the inhabitants of the metropolises. Social relations changed in the most profound way since we switched from hunter-gatherers to farmers about 12,000 to 11,000 years ago. And now we face a transformation of comparable proportions.

"Wait!" we hear readers demand. "We're surely not experiencing a revolution that impacts every conceivable aspect of social life – as industrialism did." All revolutions are, in a sense, contagions of exposure. As more areas of society become aware of what might at first seem like a pestilence or something equally as undesirable, they realize that it's being communicated, as if from one organism to another. Industrialism didn't explode: it spread. Screens also spread. What we're witnessing now is a rapid multiplication of content and use, rather than a complete break from the past. The rapidity though is probably unrivalled: the labour-saving cloth-weaving frame that became known as the Spinning Jenny and acknowledged as one of the catalysts of the Industrial Revolution was invented in 1764. But the first linear and continuous assembly process was not started until 1801 (in Portsmouth) and, while there is no definite start or end dates, the Revolution is considered to have finished by 1908, when Henry Ford began production of the Model T car in Detroit, US. Experimental broadcasting in television didn't begin until 1939 and sales of television sets spiked during the 1950s.

Trying to establish official dates for revolutions is obviously ridiculous, but the point is: in a relatively short period of time, screens have changed everyone's lives. Millennials—the generation born between 1982 and 2004—have matured in a world where looking at and running fingers across screens of some kind is not just normal but "natural." As recently as 2012, most people used their mobiles, or cell phones, for making or receiving calls. Smartphones that combined regular phone features with computer operating systems, internet access, video cameras and other digital facilities, became popular with the rise of high-speed mobile broadband known as 4G LTE. Even a cursory look at any public or private space alerts you to the speed with which we have changed: everyone seems to be fixated with their smartphones. Some of them look hypnotized.

As the industrial revolution changed our habits, relationships and mentalities, so have smartphones. How and with what effects? We're not the first people to have asked these two crucial questions. We are, however, the first to try to provide answers, not by expressing opinions in a dogmatic fashion, or by assuming that repetitive activity is the same as compulsive behaviour. Or even that, like gambling, shopping, eating abstemiously and other social habits that have been identified as addictive, engaging with screen devices is always and inevitably going to have destructive consequences on us. But by asking users themselves what they think is going on. And when we asked the questions, we did so in the most logical and appropriate manner: through their screens.

Solid as Air

Like any other project, this one started from a preconception. The textbook approach to a topic such as this is to piece together specific findings so that eventually they yield a bigger picture; social scientists call it *inductive*—bit-by-bit reasoning by inference. What researchers shouldn't do is start with a theory, hypothesis or general argument that they intend to test. But that's why textbooks are exactly that: guides to established standards of study. Metallurgists, botanists, chemists or any number of other research-oriented scholars share little with their subjects. But social researchers are very much like their subjects: so they share what the sociologist Alvin W. Gouldner (1920–1980), in his 1970 book *The Coming Crisis of Western Sociology*, called domain assumptions: beliefs about the social and natural worlds that are so ingrained that we never question them. So we can never be detached, objective or neutral; the best we can do is recognize our own biases. Gouldner called this *reflexivity* and it's a term that made its way into the research vocabulary; it means taking account of the effects of the presence of the researcher on what's being investigated. So we should own up.

We don't share the same misgivings as many scholars who study the media and its effects. Even the brief glance at recent history we have provided in this chapter inclines any rational person to suspect overreaction. It's likely that every significant shift in our patterns of behaviour is likely to arouse alarm, especially when preceded by a technological innovation. So, while there may be perils in *Screen Society*, we have opted to discover what the people who make up that society think. We all use social media ourselves and at least one of us (so far) has incurred the wrath of others sufficiently to warrant a following of trolls, at least temporarily.

But, while we've tried to approach the project as free from preconceptions as possible, there has been one possibility that has hung in the air for a while and which we incorporated into our research frame of reference. It derives paradoxically from the speculations that followed the transition from agricultural to industrial society. Why paradoxically? Because it seems on first sight a contradiction of everything we believe about the newness of *Screen Society*.

Very early in the research, it occurred to us that we were witnessing what the French theorist Émile Durkheim had, in the late nineteenth century, called *solidarité mécanique*. Translated as mechanical solidarity, this referred to the unity or agreement of feeling and action among a community of individuals who shared a common interest—usually the survival and continuance of a small-scale grouping, such as a village or town. There was a kind of small town unity in preindustrial times, social cohesion and integration developing out of homogeneity and singularity of purpose. Individual differences and conflicts were commonplace, of course; but Durkheim, in his 1893 classic text *The Division of Labour in Society* believed the tightknit network of life with similar work, education and families held people together in a community.

This coherence came under threat when industrialism gathered momentum in the late eighteenth century and the introduction of machinery made many physical forms of labour redundant. People moved from the towns to the cities, where work in factories was available. People were forced to adapt to new patterns of social life. Durkheim saw people being hurled around in a kind of particle accelerator and out of this chaos, a new social universe was created. This time it was held together, not by common bonds and a sense of unity, but by dependence: individuals relied on each other for their work and services. They might not like or even know each other, but they knew their existence probably depended on others. And vice versa. Durkheim called the attachment that emerged from this complex pattern of interdependence *solidarité organique*, or organic solidarity.

The division of labour lay at the heart of Durkheim's analysis of the transition to industrial society: as societies grew more complex and fragmented, our roles became evermore specialized. The solidarity that held humans together was like a living being, growing and acquiring a new character as it developed. The old-style community was largely overwhelmed by the industrial complex. It survived only here and there, in rural parts. Its historical destiny was secured by the industrial juggernaut.

The conventional wisdom is that mechanical solidarity was strictly a feature of preindustrial life when we were less atomized—that is, individual units who knew our neighbours but were not interested in forging close ties with our work colleagues or people beyond. Various writers came up with names, such as *One-Dimensional Man*, as Herbert Marcuse called him in 1964, who was part of what David Riesmann, in 1961, called *The Lonely Crowd*. The kind of community that brought together likeminded people with shared values and interests was considered gone.

The reader will probably now anticipate where we are headed: mechanical solidarity has been rescued from its deathbed, or, maybe even brought back from the dead. Missing presumed dead for at least a century, the solidarity that once supported and sustained communities before the onset of industrialism is actually alive and well, kept hale and hearty by people with screens. Online communities, as they are called, are not just metaphors for the real-thing-that-no-longer-exists: they are the real thing. Just because people may not live in the same place nor anywhere near each other, doesn't mean they shouldn't be considered inhabitants of the same space; it's not physical space, of course; it's cyberspace. They have characteristics in common that bring them together and help bond them into a community.

So there is a coexistence of Durkheim's two types of solidarity. They don't rival each other. They don't interfere with each other. They probably don't share much with each other either. But they do coexist, one visible in our crowded streets where strangers pass each other in silence, barely lifting their heads in acknowledgement, the other created through our perpetual interactions with others, many of whom may be thousands of miles away. And, while many will think we exaggerate the significance of this coexistence, we believe it presents one of the most momentous social developments in decades. This book is an attempt to understand those developments and anticipate some of their consequences through the thoughts and words of those who are living through them.

We have tried to write a book that isn't polemical, but which, we sense, has grown organically into exactly that: an argument that is strongly critical of existing views and disputes much of the prevailing research. We'll leave the reader to discover how, but allow us to highlight some of the areas where we take issue with convention.

We use the term Screenagers to describe people who habitually use screens in order to negotiate their daily lives. We realize that people will assume we're referring to another age bracket, like Generation X or Millennials, but we Screenagers span several different age bands: we define them by their multiple literacies rather than their ages.

Addiction is one of those words that has been stuffed with meaning like an overfull suitcase that has belts tied around it to keep the contents in. We don't mind the term being distorted, but we do mind its uncritical application to screen use. There are no screen addictions, nor addiction of any kind associated with our use of screens. We'll explain why in the pages to follow. Let's just say the form of addiction often attributed to using smartphones and other screen devices is a case of history repeating itself. Television was once thought of as a menace.

There is a widespread tendency to slap the label addiction onto any pursuit that seems to offer rewards invisible to the researcher, but which engrosses humans to an extent the researcher finds troublesome. Our objection to this will become apparent in the chapters to follow. It seems intellectually lazy to conclude there is addiction without at least exploring the deeper causes or the conditions under which supposed addiction occurs. Technology facilitates. That's all. Human beings who become addicted to practices that have adverse consequences or substances whose prolonged use has damaging effects, do so for reasons. They are often trying to escape circumstances they find intolerable, uncontrollable or repellent and, in the process, surrender some or all freewill. It's conceivable that there are people addicted to screen, though we didn't come across any in this project and, pushed for an answer, would say they are in an extremely tiny minority, but any decent researcher would want to know how the craving, habit or compulsion came into being. Screens are not addictive: social and personal situations impel humans towards the kind of enslavement implied by addiction.

Trolls are also subject to undue exaggeration. Many people assume they exist. We prefer to investigate. Our conclusions are basically that there is no self-identifying group of individuals who call themselves trolls and dedicate themselves to making life hell for other internet users. Our view is that anyone who uses the internet is capable of posting messages that are either designed to hurt others or inadvertently cause harm to others. There is trolling to be sure. The people who get hurt are often the most vulnerable too. In fact, many users believe they are the only people who get hurt. Most Screenagers just laugh at the users who try to be nasty and abusive: they're regarded by many as a joke, not a menace.

One of the more contentious areas in which digital media has made an impact is in politics and once more we find ourselves mired in myths, untruths and falsehoods. *Let's be clear.* This is a favourite phrase of politicians. They typically use it whenever they want to say nothing of importance. Let us be clear (and we do mean it): Screenagers see straight through them; the uninformed voter no longer exists. This is because they're never more than a click away from data that will help bolster their knowledge and shape their opinion. We're not suggesting they are any nearer the truth. But they're not kidding themselves about this. As television revolutionized politics, forcing any aspiring politician to become a polished entertainer, so digital media has and will continue to pressure politicians into *engaging*. This is the key term: engaging is used laxly in popular discourse. We try to be more precise in our analysis. For the moment, let's acknowledge that politics is in the throes of the kind of change not witnessed since the arrival of television.

One of the promises of the digital age is the disappearance of sexism or discrimination based on gender. It's easy to see how digitization of media could bring about significant changes in the way we understand gender and sex. In society generally, there have been breathtaking changes in our perception of gender in a relatively short period of time. Legal cases have either confirmed or afforded protection against discrimination to women and transgender persons. These have rippled through society, liberalizing attitudes and reactions. But these are not necessarily reflected in cyberspace. In fact women experience what they consider misogynistic abuse, and even blame other women for perpetuating this. Far from being a portal to an enlightened world, the internet provides only a platform for a type of inequality that seems to be vanishing everywhere else. We'll explore why a platform that offers such a bounty in so many other respects, has only paltry donations in respect of gender.

The book will also examine how our relationships to people who are key to our lives have been changed by screens. For example, doctors and lovers. We should clarify that we deal with these key people in different chapters. Digital technology is fundamentally altering the way we engage with both. In the case of lovers, we actually mean prospective lovers because the manner in which we search for, scrutinize and eventually grow to love people with whom we share our lives is being transformed. Dating in *Screen Society* is, as the reader will discover, full of contradictions. It's also risky, though perhaps not in the way we should expect.

One of the abiding themes in this book is that there is little to be worried about: scaremongering about screens is simply spreading frighteningly ominous reports that have little substance. There is just one thing we need to guard against when we engage with our screens: spending too much. We were consumers before the digital age. Now we are hyper-hyper-consumers. And those who wish to sell us products have realized it. Celebrities of all orders are paid to encourage us to buy practically anything. And they encourage us through Instagram and other digital outlets. We're not so gullible that we accept their recommendations and we're astute enough to notice blatantly commercial endorsement deals when they arrive on our screens. Screenagers remain alert to the onslaught of advertising.

Perhaps the biggest casualty of the digital world is privacy. The way we understood, respected and reacted to what used to be called a private life has now changed. Perhaps it has disappeared forever. The concluding chapter considers how our conception of privacy has changed and how this will, in turn, change us.

Research

As a researcher, you take your pick. There is an embarrassment of choice when it comes to methods. The art of research is in selecting and designing a way of approaching your subject area that allows you not only to observe, but to explain, understand and interpret. At least that's our approach and it's one we've been using for a number of years and in various contexts. In every study conducted online, the effort was to provide participants with a space in which they could express their views, elaborate their theories and share their perspectives with the canopy of anonymity and the knowledge that their thoughts would be taken seriously.

For those readers with an appetite for the technical details of how we gathered the data on which the argument of this book rests, we have included at the end of the book a section devoted entirely to revealing the nuts-and-bolts of the research project, which is based on the views of over 2000 participants. This was done across three separate phases, with phases two and three building on the data collected from the previous phases. Various methodological strategies were utilized in this project including engaging in social media, online forums and live-streamed debate as well as conducting interviews through newspapers and radio.

Such is the methodological advantage of conducting online research through surveys, participants responded from all over the world. But in order for this to happen researchers have to make the survey a valuable exercise for potential participants. One of our approaches was to include hyperlinks to news stories or research to challenge participants to think in ways that a traditional survey could not. Thus, project ownership was given to the participants; rather than impose ideas on to them, we wanted them to share with us how their lives are shaped by screens as this is what the book is about: Screenagers.

References

Alter, A. (2017). *Irresistible: The Rise of Addictive Technology and the Business of Keeping Us Hooked*. London: Penguin.

Durkheim, É. (1893). *The Division of Labour in Society*. London: Macmillan (Republished in 1984).

Gouldner, A. W. (1970). *The Coming Crisis of Western Sociology*. New York: Basic Books.

Hampton, K., Sessions Goulet, L., Rainie, L., & Purcell, K. (2011). *Report: Social Networking Sites and Our Lives*. Washington, DC: Pew Research Center. Available at: http://pewrsr.ch/2mOq4hu. Accessed March 2017.

Männikkö, N., Ruotsalainen, H., Mietunen, J., Pontes, H. M., & Käärläinen, M. (2017, December 1). Problematic gaming behaviour and health-related outcomes: A systematic review and meta-analysis. *Journal of Health Psychology*, 1–15.

Marcuse, H. (1964). *One-Dimensional Man: Studies in the Ideology of Advanced Industrial Society*. New York: Beacon Books.

Postman, N. (1985). *Amusing Ourselves to Death: Public Discourse in the Age of Show Business*. London: Penguin Books.

Riesman, D. (1961). *The Lonely Crowd: A Study of the Changing American Character*. New Haven, CT: Yale University Press.

Telegraph Reporters. (2017, February 7). Facebook is a 'tool for evil', says judge as mother trolled over fake claims she tried to kill a baby is found dead. *Telegraph*. Available at: http://bit.ly/2lrDvYb. Accessed February 2017.

2

History

Magic

In 1696, an awestruck English observer witnessed the demonstration of a piece of equipment that looked like a box fashioned from wood and metal and with a tube protruding from one of its sides. The tube sent out a beam of light that, on reaching a nearby wall, changed shape and revealed a series of scary images. The observer described the box as "A Magic Lanthorn, a certain small Optical Macheen, that shows by a gloomy Light upon a white Wall, Spectres and Monsters so hideous, that he who knows not the Secret, believes it to be performed by Magic Art."

The "secret" wasn't actually much of a secret. If it was, it had survived several millennia. The "optical macheen" was based on the *camera obscura* effect, which had been used since before the Christian era and was, basically, a way of capturing images through means of a pinhole and an external source of light. It was used for tracing and drawing objects, probably on parchment or perhaps animal hide. Even without technology, a blank wall could provide the surface for shadow shows or shadow theatre, in which the silhouettes of puppets or live people were projected. Dexterous performers used their hands to make shapes.

© The Author(s) 2018
E. Cashmore et al., *Screen Society*, https://doi.org/10.1007/978-3-319-68164-1_2

The history of the piece of technology that eventually became the magic lantern is hazy. It's probable that a primitive version was around in the 1400s and, over the next two centuries, it continued to develop. A German priest named Athanasius Kircher is popularly credited with making a breakthrough, though scientific studies of the behaviour of light (i.e. optics) and the development of lens with particular properties for deflecting and radiating light provided the stimulus to create what became known as the Magic Lanthorn. This was popular in Europe and Asia. As our observer suspected, those viewers who were not familiar with the technology could have been forgiven for thinking the images of grotesque-looking demons were conjured by means of magic. Hence the name of the device.

There was no magic, of course: just a simple form of image projection for showing drawings and, later, photographs. In 1839, Louis-Jacques-Mandé Daguerre, presented the first practical photographic process to the French Academy of Sciences. Daguerreotypes, as the images produced by the process were known, were transferred to plates, or slides, and the pictures were enlarged and shone onto blank surfaces. The light necessary to illuminate the images was provided by oil lamps or limelight (intense white light obtained by heating lime) and, later gas, but more reliably, from the 1880s, by electricity.

Writing in the publication *International Bulletin of Missionary Research*, Donald Simpson, in 1997, recorded how, in the late nineteenth century, British church missionaries were enthusiastic users of magic lanterns (as lanthorns were re-spelled) when promoting Christianity overseas. Schools also used magic lanterns to illustrate lessons. But the appliance was more than an instrument of instruction: it was a means of entertainment. Meredith A. Bak, in 2015, wrote of a parallel history of the magic lantern, at first as a tool-of-the-trade of travelling showmen who toured in both the UK and US in the 1800s, amusing paying customers with what must have seemed astounding images, some drawn, some photographically reproduced; and later as a children's toy. Remember: audiences were not familiar with any other method of screening images, so the sight of magnified representations of exotic beasts or grotesque monsters as well as more recognizable figures must have been stunning and perhaps bewildering. And probably entertaining, though there was no entertainment industry in the way we understand it today and there was no specific pursuit or practice marked out as entertainment. The showmen, who were known collectively as the galantee, would provide fables or moral tales with pictures to embellish the stories. The element of instruction or teaching would have been present.

Phantasmagoria were displays of images on screens or sometimes smoke that seemed like dream sequences. They were designed to frighten not educate audiences. The most celebrated Phantasmagoria impresario was Etienne Gaspard Robertson, who employed several portable magic lanterns mounted on wheels to give the illusion that the images were moving. It's probable that there were moral fables in the exhibitions, but the main effect seems to have been to terrify audiences with grotesque pictures. Perhaps they were intended as warnings to those who strayed from the path of righteousness. Whatever their purpose, being scared stiff appears to have been what audiences in nineteenth century Europe desired (horror films suggest many of us still like the thrill).

While itinerant showmen continued to ply their trade, some permanent venues became noted for their spectacular and elaborate magic lantern displays. London's Royal Polytechnic Institution, for example, installed several screens of eight metres, or 25 feet diameter. The installation incorporated several projectors and musical accompaniment. It was founded in 1838 and over the next several decades, "the lantern's association with traveling showmen was severed, replaced by the precision, widespread availability and technological sophistication of the industrially produced scientific lantern," writes Bak in 2015 (p. 112). At the Polytechnic, the magic lantern's utility as an educational tool was fully realized.

In 1895, a new and altogether more exciting version of screened display was exhibited in Paris. Two brothers, Auguste and Louis Lumière had been experimenting with a version of an earlier machine called the kinetoscope, which allowed viewers to view moving images through a peephole. The premise of the invention was to transfer a series of photographic images on a film then to run the film across light so that the rapid passage of the individual images gave the appearance of motion. The Lumière brothers took the idea of motion pictures but combined it with the projection technique of magic lanterns. The moving images appeared on screens that could be watched by hundreds of people.

It was a short step from perfecting the display to generating fresh content using new photographic images and the combination, as the reader will guess, gave us films. Within twenty years of the Lumières' launch, film crews were using lighting, stage sets and other effects to recreate naturalistic settings. Because the motion pictures had no sound, the noise of moving around heavy cameras and directors issuing instructions did not interfere.

The magic lantern must have seemed abysmally unsophisticated by comparison, though Bak argues persuasively that it "was instrumental in formulating a distinct culture of domestic media spectatorship and particularly in training children to understand visual entertainment as a repeatable consumable form of leisure activity" (p. 113). So, by the time, motion pictures were circulating, audiences had been inured to the practice of gathering in relatively large halls, or auditoria, and watching respectfully mute, so as not to distract other consumers. Bak refers to the "self-regulation" of audiences: in bars, taverns and other places of public resort, people might have been unruly, disorderly and unrestrained; in film theatres, they behaved with decorum.

Silent movies were also the catalyst of what we now call entertainment. This sounds an extraordinary claim, but consider the concept of a specific portion of people's time when they could engage in a pursuit that wasn't intended to improve them morally or instruct them in some practical way and that was reserved specifically and exclusively for enjoyment, diversion, relaxation and perhaps for restoring the spirits after a day's or a week's work is probably newer than we typically assume. We know that drinking hostelries, such as inns and taverns, were around in preindustrial times and may even date back to medieval England, possibly as early as the 1400s, but we don't know their purpose: restorative perhaps. Think of the word recreation: to create again, or renew, perhaps in a physical and mental sense. Entertainment may have existed before the industrial revolution, beginning in the late 1700s, but it didn't morph into a freestanding activity concerned with the provision of organized activities designed for the recreation of industrial workers until the nineteenth century.

The British theatres known as music halls began appearing in the 1850s and these offered a mix of songs, comedy, conjuring and speciality acts. So, when cinema appeared as a rival, there was a ready-made template for entrepreneurs and consumers alike: a spacious theatre with seats, a stage and an understanding that patrons would respect others' comfort. And, of course, a box office: this was the kiosk where admission tickets could be bought, or reserved in advance. As well as being prepared to sit peacefully in relative silence for long periods staring at a finite space, theatre audiences were familiar with the idea of parting with money in exchange for a presentation that would engage them agreeably. This exchange agreement was the nexus of what became the entertainment industry. So, it is only appropriate that we now move to Hollywood.

Stars

Try watching those silent films from the 1920s. Like *The Kid* from 1921, in which Charlie Chaplin had audiences crying with laughter. Or the 1925 adaptation of Gaston Leroux's novel *The Phantom of the Opera*, which was one of the first horror films. Fritz Lang's colossally ambitious 1927 epic *Metropolis* inspired practically every futuristic, dystopian and cyborg movie ever made. Audiences were understandably overwhelmed by these silent classics: they watched with fear and wonder, the only sound being the piano accompaniment in the theatres. For many, the pre-sound era provided some of the most beautiful and inventive films ever made. There were many more films that kept audiences rapt in the 1920s and have been retrospectively considered classics of the era. But, we repeat the invitation: try watching them. Interesting and intellectually absorbing as they are, they don't quite pass muster as entertainment, at least not in the way we understand it today. *Metropolis* has been remastered many times with various soundtracks and still plays in theatres, though audiences watch reverentially, in admiration and respect.

We may enjoy them, but not in the same way as we enjoy films today. It would be slightly surprising if we did, if only because our sensibilities change over time. This means the quality of being able to appreciate and respond to complex aesthetic and emotional stimuli never remains the same: we—humans—are liable to be excited, offended, shocked, pleasantly amused and sentimentally moved by all kinds of things; but those things change over time. The quote with which we started this chapter indicates how the effects of what now seems a crude and childish device cast a spell over audiences. Imagine how the first viewers of silent movies, just a couple of hundred years later, would have responded. Early movies must have seemed more staggeringly different to the old magic lantern shows than the silent films are to modern motion pictures. And yet there is a common denominator: audiences were all gazing at a fixed two-dimensional surface that seemed to come to life when light projected images onto them.

The First World War ended in 1918. By that time, a film industry had emerged in Hollywood, a district of Los Angeles in California. The industry had been started mainly by Jewish migrants from Europe: anti-Semitism prevented many from getting jobs, especially in trades controlled by unions, so they opted to start their own businesses in the fledgling entertainment.

"Talkies," as the films with sound were called, did not eclipse silent movies: for years, the two forms coexisted. It seems strange to think audiences remained rapt by silent films when they could watch and listen to the new types that became available from 1927 when *The Jazz Singer* was released. Film theatres were set up for silent films, of course; so the introduction of sound forced the owners into either re-equipping cinemas (at some cost), or keeping their fingers crossed while hoping the new talkies would fizzle out. The opposite happened and silent film was soon history. Cinemas were revamped and the era of talking pictures began, heralding what became known as the Golden Age of Hollywood.

By 1930, film had become a mainstream recreation. Eight major film industries surfaced, each churning out enough films to keep consumers returning to film theatres every week. Remember: there was no television at the time, so the main rival to film was radio. In terms of cost and convenience, radio won: it was relatively affordable and involved no travelling. But it must have seemed an anaemic form of fun alongside film.

The film industry made an important discovery too: audiences flocked to theatres to see particular actors as much as (if not more than) they did films. The silent era had occasional stars, like Chaplin, Greta Garbo and Lillian Gish, but actors were, for the most part, contracted employees who worked for the studios without any expectation that they would enhance the box office appeal of a film. Talking pictures, on the other hand, humanized many actors. Audiences could hear as well as see them and the industry sensed pulling power in the actors such as Mae West, Fred Astaire and the Marx Brothers. They could also evaluate them on their acting rather than just looks. Whereas a silent screen icon like Rudolph Valentino might have lacked technique, application and discipline, he could get away with looks alone. But sound introduced new types of figures. One of the studios, MGM reportedly boasted that it had "more stars than there are in the heavens." The name stuck.

Films were black-and-white and the sound quality of the talkies was poor. Box office fluctuated as the industrial and financial slump known as the Great Depression took effect. But filmgoers continued to pay their money, inciting studios to experiment in new technologies to maintain interest. Of the many innovations in the period was Technicolor, which was a process of colour cinematography using synchronized monochrome films, each of a different colour. The result was a full colour film. Early attempts were mediocre, but, in 1939, *The Wizard of Oz* and *Gone With the Wind* were both artistic and box office triumphs that sealed the future for Technicolor films.

Among the other technological innovations that met with less success was 3D, which was popular for a while in the 1950s but faded only to return in eye-goggling depth in the early twenty-first century. Smell-O-Vision was one of those ideas that probably looked good on paper. *Femme fatale* appears on screen and audiences smell Chanel No. 5. Maigret, the pipe-smoking detective, comes into view and audiences get a whiff of burning tobacco. Bunch of teenagers meet at a pizza restaurant and audiences smell … readers will get the idea. The technique was trialled in a 1960 film called *Scent of Mystery* (honestly) but it never took off. Sensurround added massive low-range speakers to movie theatres screening the film *Earthquake* in 1974. 4D, as film enthusiasts will know, is one of the more recent attempts to create an immersive experience for audiences. None of these has been revolutionary. The jaw dropping developments that have mesmerized consumers have been on the screens: the size of them, the colour and the detail that became clearer as physical 35 mm celluloid film was replaced by high resolution digital technology; and the gigantism offered by huge IMAX screens, some as tall as football pitches are long.

It might seem a ridiculous question but would all these technological developments in film have happened without television? In some measure, film reacted to the threat offered by tv in the 1950s by outdoing it. All the big innovations, successful and unsuccessful, were attempts to wrest paying customers away from the comforts of their homes. As a matter of fact, cinema has been engaged in this kind of battle ever since. It has been constantly challenged to come up with attractions that can't be experienced in any environment apart from the theatre. But, at one stage, television vs. cinema looked like a case of David and Goliath, with the smaller combatant all set to upset the bigger, but cumbersome opponent. Goliath, in the form of film, watched helplessly as box office receipts fell.

Television: Audiences and Shoppers

Television was, in many ways, a straightforward extension of radio. In both the UK and US, broadcasting organizations that operated radio were the natural custodians of the new experimental service. Radiovision and radioscope were among the many working names prototypes were called in the early twentieth century. So it seemed logical that television (as the name was settled) would be operated and governed by the same people who managed radio. Britain's British Broadcasting Corporation (BBC) was mindful of its ancestry and mission: it had been established by royal charter in 1927 and held a monopoly on the condition that it remained impartial in its reporting and free from advertising. The BBC's mission was: "To enrich people's lives with programmes and services that inform, educate and entertain." The placing of the words in this order was probably deliberate.

American television had no such remit: its job was to advertise and the programmes were bait to get viewers to their tv sets. This sounds a crude appraisal, but radio worked on similar terms: the shows were sponsored by the likes of Macy's department stores, Metropolitan Insurance and American Express. Even if some radio stations in the early 1920s bought into the BBC model, they were soon reminded that advertising revenue was their lifeline. British radio, by contrast, was funded by the government through a licence fee.

After the Second World War, when television emerged as a warrant-able addition—not a rival—to radio, there seemed no obvious reason not to let radio steer the new mass medium. Remember: no one could predict whether or not television would succeed at all, let alone become arguably the defining cultural innovation of the century. So commercial organizations were not falling over each other to invest in what might easily have gone the same way as N-ray, which was supposed to improve on X-Rays, or the Otto Lilienthal glider that was soon surpassed by the Wright brothers' contribution to aeronautics. The prospect of organizing a night out at the theatre and spending three or more hours in the company of other fascinated cinema fans, then discussing the films (there were usually two features) seemed altogether more enticing than staying at home watching a comparatively tiny screen flickering with hazy images—reception in the 1950s was often poor.

Television seemed contrived, anyway. The BBC's programmes were influenced by the aims of its first general manager, John Reith, the son of a Presbyterian minister, who believed the Corporation was or at least should be a moral force. American programmes, by contrast, were influenced by advertising imperatives: if a programme drew the right kind of viewers to their screens, then it was a success; if it didn't it was dropped. Serialized domestic dramas spread over periods of time were especially popular and were sponsored by companies such as Procter & Gamble and Colgate-Palmolive, both of which sold household cleansers; hence the name soap operas, or soaps, as they are now known.

Different controllers with different motives. But the results were the same: consumers loved the idea of watching a screen for hours at a time, in the company of friends or family and able to eat and drink and, actually, do anything they wished while watching the shows. Sales of domestically owned or rented tv sets shot up dramatically in the 1950s: at the start of the decade hardly anyone had a tv, but, by the end, over 90% of the population of the UK and US had at least one set. By this stage, the novelty of owning or having unrestricted access to a device that, while not a functional equivalent of cinema, was emerging as a rival, had worn off. Replacing it was idea that tv was indispensable: "every home should have one," the advertising slogan of the day suggested.

For a while, the film industry seemed in sharp terminal decline, though, as we know, it recovered and still thrives. Television, meanwhile, ascended vertiginously: it was fresh, occasionally spectacular and unafraid to experiment with new forms. The soaps were one, quiz shows were another, in-depth documentaries and regular news bulletins were others. The cultural landscape didn't just change: it seemed to have been shaken by an earth tremor. It's difficult to think of any aspect of social life that wasn't affected by television. Not just our daily habits, but our relationships with others, our awareness of world events, our reaction to politics; the effects of television were many and multifaceted.

In the early twentieth century, American radio complemented billboards (or hoardings as British call them), those large outdoor boards displaying advertisements. These and newspaper advertisements were the two media used to thrill potential consumers into buying commodities. At best, ads would provoke feelings of desire so great that a person would do practically anything, including breaking the law, in efforts to satisfy his or her craving. At least this is how the sociologist Robert K. Merton made sense of crime: he wrote an article in 1938, "Social structure and anomie."

Merton concluded that the ultimate goal for consumers was material success, which they wanted to display—and display *conspicuously*. Good clothes, cars, electrical appliances: these were all commodities that were relatively recent arrivals in the marketplace and ones that people wanted. People valued their ability to consume and they were encouraged through advertising, to maximize this ability—within certain boundaries. Merton's view was that the boundaries defined the legitimate means through which people could achieve their goals. There are right ways and wrong ways to achieve them. When people strove for material goods but lacked the means to get them, they often opted for the wrong ways. In other words, they stole the goods that the posters and the radio commercials were telling them they should have.

The "non-conforming" conduct, as Merton called it, was a response to this condition.

It's tempting to blame television for intensifying this tendency and, of course, it was blamed for many other socially undesirable tendencies. In 1951, the Code of Practices for Television Broadcasters stipulated generous limits on the amount of commercials that could be shown—for example, an average of 13-minutes per hour.

In Britain, commercial television arrived in 1955, with ITV. The BBC, remember, was not allowed to screen advertising (and still isn't). Like its American counterparts, ITV advertised products that promoted a conception of the Good Life: materialistic and full of appealing material possessions and automatic labour-saving appliances—washing machines, vacuum cleaners etc.—that would free up time to enjoy leisure.

Audiences are also shoppers: the media have been treating them as such since the early twentieth century and no one has resisted this style of management. We choose the term with care: media, whether traditional or digital, manage us. We're still free to choose what programmes we watch, where and when we watch them and audience freedoms have expanded rather than contracted since 2007 when Netflix developed its own technology to power its pioneering streaming service that delivered films directly to computers. But, advertising ambushes us at every turn. You can't engage with media without encountering ads. And, in a way, we have television to thank or blame for that. It inured us to advertising. No one today would expect the 120-minute film they'd paid to watch at the cinema interrupted every fifteen or so minutes. But the quid pro quo offered by commercial tv was that viewers surrendered their attention to advertisers in exchange for entertaining programmes. (Film's solution to this irksome problem was product placement i.e. featuring branded products in the movies.)

We'll discuss consumerism and advertising in more detail in Chapter 12, but for now we want to remind readers that our tolerance to advertising is a product of television. We learned how to enjoy the experience of watching drama, comedy and news even with the persistent, unwanted and often exasperating intrusions of advertisers. This kind of zigzagging between welcome and unwelcome content would be unthinkable in many spheres: theatre, concerts, ballet, education and many sports are among the activities that could theoretically be broken into segments to accommodate advertising. Yet somehow, we think there would be resistance. With tv, there was none and, in this sense, it cleared the way for all screens.

Advertising is one of the ties of kinship that makes television the closest relative to today's screen devices. There are other similarities. Like computers and phones, a tv is capable of being owned. No one owns a cinema; we visit them and take advantage of their facilities. But tv's are usually our own and, in this sense, they were like the things their programmes advertised; in other words, possessions.

In later chapters, we will evaluate some of the arguments about the supposedly abominable consequences of our involvement with digital media. So we should point out at this early stage that most of them have already been associated with television. Concerns over the effects tv was apparently having on audiences were commonplace in the 1960s. Watching the box was thought to have a dire impact on our cognitive abilities, specifically by shortening our attention spans. Television's detrimental effects on traditional family life were seemingly obvious as viewers sat around with their eyes fixed on the screen instead of talking to each other. And what of actual behaviour? All that violence on the screen was sure to translate into actual violent behaviour as we became desensitized. How did such a dysfunctional piece of kit withstand the criticism? The answer is simple: by providing so much pleasure.

Having established itself as a central part of life in the 1960s television has tenaciously stayed there, morphing into a partner of internet couriered services, like Netflix and Amazon Prime, without seeming to overstay its welcome. Television has a suppleness that critics didn't understand when, in the early twenty-first century, they predicted its demise. Even today, television can't seem to make up its mind whether it still intends to enrich our lives with art, or just entertain us. It does both, of course; and, at the same time, contrives to keep us shopping.

Magic lanterns might have been the ghostly templates for screened edification and entertainment, but television is the archetype for today's screens. Every so often, there are murmurings about how out-of-place tv has become. In 2014, for example, the *Telegraph*'s Christopher Williams observed "How young viewers are abandoning television" and, in the same year, the *Financial Times*' Jonathan Ford wondered "Is YouTube the new television?" But television has adapted, survived and, indeed, prospered; it was and is the life force of *Screen Society*.

References

Bak, M. A. (2015). 'Ten Dollar's Worth of Fun': The obscured history of the toy magic lantern and early children's media spectatorship. *Film History, 27*(1), 111–134.

Ford, J. (2014, November 14). Is YouTube the new television? *Financial Times*.

Merton, R. K. (1938). Social structure and anomie. *American Sociological Review, 3*(5), 672–682.

Simpson, D. (1997). Missions and the magic lantern. *International Bulletin of Missionary Research, 21*(1), 13.

Williams, C. (2014, October 8). How young viewers are abandoning television. *Telegraph*. Available at: http://bit.ly/2mQddMI. Accessed March 2017.

3

Screenagers

New Literacies

Some time ago, one of the present authors was presenting a lecture to a class of undergraduate students, one of whom appeared perpetually distracted by a smartphone on his desk. "Do you see any value in the theory I'm discussing?" the lecturer directed a question, not expecting much of an answer, but the student didn't hesitate:

> Well, the Uses and Gratifications model was a significant advance on earlier theories of media in which audiences were assumed to be passive and subject to what Lazarsfeld called 'narcotizing.' In the 1970s, Blumer and Katz proposed that audiences were active and used media content to satisfy various needs. So, I'd say it was valuable.

In other words, a pretty decent answer that indicated that, far from being distracted, the student was paying attention—far closer attention than the lecturer assumed. It was a useful reminder that comprehension is not just a straightforward ability to understand something: it's more like a collection of abilities that enable us to understand many things and through many different modes. Those modes include traditional

© The Author(s) 2018
E. Cashmore et al., *Screen Society*, https://doi.org/10.1007/978-3-319-68164-1_3

written material, literacy being the ability to read and write. But there are now different types of literacies, such as computer literacy and, more generically, digital literacy, which means processing data from several digital sources. Reading from paper and off screen, listening to information in the room and through ear pods and forming ideas through conversations face-to-face and online: these are some of the ways and conduits through which we receive and interpret data today. It's why we often use the word multimodal to characterize several different modes of activity. Listening to a lecture and reading online is a piece of cake to *Screenagers*.

We use the term to describe a generation of people that use their screens "many times per day." That's about 80% of us. Only 18% describe their screen seeking habit as "several times per day," 1.4% consulted their screen once per day, 0.4% "less than once per week." Unlike previous generations that are defined by years of birth, Screenagers tend to span the years. Let's briefly recap how other generations have been depicted.

Baby Boomers: This is the label given to people born in the years following the end of World War II. There was a sharp increase in the birth rate just after peacetime resumed. Although there's no official cut-off year, the Baby Boomer generation is generally thought to have ended around 1964, after which there was a break before another distinct cohort emerged.

Generation X, or just *GenX*: Members were born in the two decades before 1984. This is a very rough approximation because no one in the late 1960s or early 1970s was using the term Generation X—unless they were referring to Billy Idol's band, formed in 1976 (and named after an earlier booklet about 1960s youth in England).

Millenials: These were born in the 1980s and 1990s, as we approached the end of the twentieth century. Hence the name millenials. There's no official categorization or birthdates, of course.

2000 onwards? Nomenclaturists must have become bored, then given up devising names for generations. If you were born this century, you don't have a named cohort. Yet it strikes us that this group have a distinct and evident identity, though not one that's strictly demarcated. We mean that, in common with most other people today, they use screen devices for manifold reasons: not just accessing news, emailing, or messaging; but for ordering food, hailing cabs, getting directions, turning on the central heating and countless other tasks that no one ten years ago imagined could be done on a single appliance.

Screenagers are not so much defined by year of birth; more by *critical literacies*. This is a term coined by Michael F. Graves and his co-authors of the 2011 book *Teaching Reading in the 21st Century: Motivating All Learners*. Their point is that reading, in the widest possible sense, incorporates much more than reading sentences on a page: it involves, among other things, phonemic awareness, word recognition, vocabulary, fluency and playing interactive reading games. And we should probably add: listening to lectures—probably at the same time. This is not multitasking in the conventional sense (doing more than one job at once): it is expanding the capacities to communicate and learn via different media simultaneously.

'Multimodality' is a term used by linguists and communication theorists to describe the way in which we employ various channels simultaneously, though Rozália Klára Bakó, of the Hungarian University of Transylvania, in Romania, observes how the term has been redefined "by information technology researchers to express this variety enhanced by the interactive World Wide Web." Bakó's point is that today, we have all become multimodal, if only because of the sheer fact that we are alive. Navigating our way through a day's work necessitates being able to operate several modes of communication.

Before we run the risk of exaggerating Screenagers' abilities so that they appear to be super-brains, we want to be clear: learning, or knowledge acquisition is not a fixed capacity any more than any of our physical capabilities are finite. Somehow we find ways of extending them. Imagine if we had told someone in 1960, when the men's world record for 100 meters was 10 seconds that, within 50 years, an athlete would run the distance in 9.58 seconds. They would have dismissed the possibility as laughable. But Usain Bolt, as we know, ran that time in 2009. Our learning capacities don't change in a similarly uniform way; but, like every other aspect of our being, they adapt to changing environments. Our environment today is dominated by technologies. Even people who proudly proclaim themselves as technophobes with no interest in modern technologies have to engage with technology, if only because their survival depends on it.

In 1983, Robert MacNeil wrote an article for the *New York University Education Quarterly*, in which he laid out a question in his title "Is television shortening our attention span?" It was one of a number of measured analyses that followed the hysterical scaremongering of the previous two decades. Neil Postman's 1985 book *Amusing Ourselves to Death: Public Discourse in the Age of Show Business* was perhaps the most intellectually robust denunciation of television. Postman believed television to be responsible for, among other things, "our brief attention span and our appetite for feel-good content," which, as he put it, "can short-circuit any meaningful discourse." MacNeil's view concurred with Postman's. Neither accepted that our mental abilities adapt in accordance with technologies. We're not determined by the technology we use: we learn to adapt to and shape it to our requirements. As we change our environment, we change ourselves, both cognitively and physically.

"If you can't use a computer it's the equivalent to my generation not being able to read or write," one of our participants believed, suggesting how we survive as a species by rising to the new demands of critical literacy. Questioned specifically about these demands, respondents recognized interactivity as a key difference. One observer compared the literacy demands of television with what she called the "proactive" demands of other screen technologies: "TV is reactive. You watch what it shows. [With the] internet you can find what you want. It's more proactive." And another: "Online activities are usually more interactive than television, which can only be a good thing." A good thing?

Digital + Real Entangled

The interactive web 2.0 platform emerged in 2004–2005. No one, or at least only a very small minority of people, predicted its effects. Many probably realized its technological potential. But not its potential to change us, the users. Basically, it made possible interactive communication between users, not just of computers, but of various portable devices, including phones. Wherever you are, look up and you will not have to wait long before you see a device with interactive potential. You may even be using one while you read this; if not you can probably see someone who is. Prior to 2.0, we used phones for conversations and exchanging messages; we used computers for accessing information and emailing. Then the vectors (i.e. the directions, or routes) of communication diversified. Some technological innovations leave us unmoved. 3-D television? Electric self-balancing unicycles? Others challenge us to master them. Web 2.0 did this.

"Who knows what this will unleash?" wondered a participant in the *Screen Society* project. He was contemplating the impact of interactive media on future generations. Others were more secure: "I believe that children need online socialisation as their future lives will be lived through many online contacts." It's possible that the second remark will strike as much fear as the first. "Lived through" is one of those phrases that immediately conjure images of a world in which we are somehow dehumanized by media technologies so that our existence is limited. So, when scholars such as J. Sage Elwell, of Texas Christian University, observe, "The line between life online and life off-line has become blurred," it could secrete prescience or menace.

Writing in 2014, Elwell believes three developments have changed us: (1) there's been an integration of our "off-line identities" and our "online selves": he means that a few years ago we would "go online" and sometimes assume different identities during the period we spent on the net; now there is no meaningful distinction; (2) the availability of compact, portable devices means that we have the ability to be connected with multiple others without interruption, no matter what else we happen to be doing; and (3) computing is now so ubiquitous that we are surrounded by other people who are, in some way, communicating with others; lift your head now and this is obvious—practically everyone you can see will be communicating with others. "These technologies are integrated into the habitats of our lives so that they become invisible," writes Elwell (p. 235).

Some of the communication may closely resemble traditional forms: shopping online, for example, transposes browsing in a shop then purchasing. Other communications involve an altogether different mode, as a young woman in the *Screen Society* study explained:

> The language we use is completely different, from inside jokes to memes to the way we type our words in order to show tone. These interactions shape a full new language – although most natural internet users know

the language they struggle to explain it. It's not like French or Chinese, it's our everyday language being used in different ways to allow clear communication and for relationships to be formed.

What this contributor is effectively saying is that Screenagers are not simply communicating in a secret code or an arcane language: they're constructing a kind of discourse. The strength of Elwell's argument lies in his preparedness to understand this as part of a genuinely new type of life. It sounds like hyperbole and it certainly takes some digesting, but the implication is that we are witnessing the emergence of what he calls a "transmediated self," which incorporates both online and offline presences. Elwell writes: "To speak of the two as somehow separate and discrete fails to appreciate the extent to which the two are entangled and the reality that we live inside that gap—we live in the uncanny valley between the digital and the real" (pp. 242–243). (We appreciate that we're in danger of opening up a Pandora's Box when we allow "real" as an alternative to digital; many might properly insist digital is a real series of the digits 0 and 1. So we will stick to a tried-and-tested definition of *real* from Peter L. Berger and Thomas Luckmann's 1967 book *The Social Construction of Reality*: "Something that cannot be wished away," and hope this suffices for the present volume.)

Some of this is tough to penetrate, while other parts can be easily understood. As another of our research respondents put it:

> The potential for anonymity has made a far greater difference. I wouldn't say many of the things I say on the internet face to face, this includes criticism of what I regard as foolish comments when I would feel very uncomfortable doing so particularly to a female in person.

The virtue of anonymity was picked up by another participant:

> Online conversations offer an anonymity that often makes people say/do things they would not find themselves doing. This also breeds confidence in those who may not find any comfort in talking to people when they can gauge the immediate reaction of the person.

Of course, anonymity also introduces more sinister possibilities, but we will come to these later in the book; for now, we want to identify the character of the generation we've called Screenagers. Let's return to Elwell's point about "the digital and the real." "Some are different relationships, whereas some build on real life relationships," replied one participant when asked about the nature of online interaction. It's hardly precise, but the indefinite quality is characteristic of Screenagers and, in a sense, it backs up Elwell's argument about how the two are "entangled."

Some Screenagers who enthusiastically live online acknowledge: "Online relationships require a far greater level of trust [than face-to-face relationships]." Others say they are "more transient or ephemeral with no actual real feeling or compassion." This contributor noticed how the nexus between online and offline communication is "seamless":

> Online contact can be picked up and put down, allowing exchanges to take place over hours/days rather than being confined to the timescale of a physical meeting. Beyond the confines of face-to-face conversation, online exchanges can link seamlessly to other media, reports, video etc., enhancing the experience.

This is a thoughtful précis because it captures what we might call the art of being a Screenager: seamlessly navigating from one mode, or modality, to another, often severally at one time, sometimes with other human beings, at other times with different kinds of inanimate entities. It will sound to some like pseudo-depth to justify fake relationships and some of our study's participants remain cynical. This participant's summary is worth reproducing in full:

> Online friendship is one dimensional. Talking to a friend face to face is completely different to talking online. Communication does not consist of simply words but of facial expression, tone of voice, posture etc. etc. It is sometimes difficult to ascertain a person's true meaning without talking face to face. So, online friendship can never come close to offline. Imagine cultivating a long online relationship and then coming face to face with the person. It would be very interesting and then perhaps disappointing or surprising.

It comes from a more mature man and contrasts almost perfectly with the view of most Screenagers. But is also reveals the chasm between those who habitually use and, in a way, live through their screens and those who see them as helpless dupes who believe they are in a parallel universe but are in fact just deluding themselves.

One-Dimensional?

The internet has provided "young people with new spaces for identity experimentation that are unmediated by adults," as Andrea Flores and Carrie James put it in their study "Morality and ethnics behind the screen: Young people's perspectives on digital life." The subtitle of the paper suggests how the American authors approached their study. In their view, the whole point of online interaction is that it is not filtered by people who wouldn't be simpatico—in other words, adults with whom they don't share a bond. Of course, the Flores and James study doesn't appreciate that it's not just young people who engage in digital life. It was published in 2012 and the demographics of internet use have changed and are still changing. The *Screen Society* research indicates a broad spectrum of users, from children to people in their seventies. So, unlike the other generations we wrote of earlier, Screenagers have more porous age boundaries: there's no indication that people are cutting down, less still dropping their online practices as they grow older. Within a few years, we'll all be Screenagers. But let's return to the American study, which provides a counterbalance to the argument that online friendships are, as our participant called them, one-dimensional.

Flores and James begin from an observation many overlook: "Young people operate both in smaller communities of interpersonal relations *and* in large networked publics" (p. 838). This means that they're constantly shifting between spheres, such as Facebook or Instagram where they have close acquaintances, and more public arenas where they interact with those with shared interests in, for instance, World of Warcraft or Ravelry, the knitting and crocheting circle. Even this separation of spheres is artificial: like much life on the internet, there is permeability.

Flores and James believe the "networked, public nature of the internet requires a capacity for thinking more abstractly, about the effects of one's actions on unknown others and at the level of the community" (p. 838).

The study persuasively argues that interacting on the net challenges users to think in different ways, make decisions that will affect others, and choices that may have consequences; they, perforce, acquire capacities for ethical thinking. Users are stretched both intellectually but morally; they learn to view the world through the eyes of others, adopting the position of other users. In other words, the study disclosed what we might call an inner world that's typically not available to outsiders. To dismiss online relationships as superficial and poor substitutes for face-to-face equivalents misses a dimension.

Time and again, contributors to the *Screen Society* study described how communicating online lacked the spontaneity or sheer randomness of everyday physical interactions, though no one specified whether they felt this to be good or bad. But it does echo Flores and James' findings. "You can be more considered with your messages, crafting exactly what you want to say and taking your time, unlike face to face discussions," said one participant, suggesting that she strives to consider the implications of comments before hitting "send."

Several other participants made similar points, one in particular imagining how recipients of messages sent in jest can easily interpret them seriously and become offended. Without facial gestures to accompany the utterance there is always openness to interpretation. So some forethought, anticipation and even calculation typically precedes online exchanges. This seems to buttress Flores and James' argument about the moral reasoning involved in online interactions. Another complementary argument is: "Online relationships require a far greater level of trust."

Again the moral elements of friendship with someone physically removed is pertinent: contrary to popular views, Screenagers do not stumble heedlessly into friendships. And yet, "you tend to know more details about their lives but less at the same time." It's an acknowledgement more than an admission: online friendship promotes reciprocity of perspectives, but not necessarily intimate knowledge.

This rapport-without-intimacy means users "can discuss things you wouldn't face to face, with people you wouldn't normally come into contact with." So there is a closeness online that wouldn't be possible F2F (to use the popular abbreviation). Karen Zilberstein's research complements this. In 2015, she criticized doubters with her finding: "Relationships facilitated by technology appear to provide the same sense of wellbeing as is associated with other types of close associations" (p. 152). The rapport is strengthened by the obvious fact that online friendships are founded on symmetry of interests. As one participant put it: "[Online] friends share a common love or loathing."

But some would argue, if you are a vulnerable and unstable young person trying to find new friends and establish meaningful, rewarding relations online, you should ignore the foregoing passages. For all the bounties seemingly available, we should stay mindful of the earlier caution, "Who knows what this will unleash?" We know that many users emphasize the rewards, but we also realize that the anonymity afforded by the internet and which facilitates frankness and affinity provides a sheltering canopy for those with malfeasance in mind.

We will address trolling and other online mischief and intimidation in Chapter 7; for now, we want to identify the principal features and qualities that make Screenagers distinct. So far, we've recognized the ease with which Screenagers can discourse, sometimes at great length and on unrelated topics, in various modes; the multimodal communication, in which users exchange views, share ideas, view images and consider other types of data, is a key characteristic. Sometimes dismissed as inarticulate, Screenagers are able in multiple literacies.

New and old worlds, or the digital and the real have become increasingly entangled, to use J. Sage Elwell's term again. And Screenagers are able to slide comfortably between the two, negotiating multiple identities or selves as they conduct their affairs. This is difficult, if not impossible, to identify from the viewpoint of an outsider critic; it becomes evident only through the responses of users themselves.

Critics see insults and danger everywhere on the net. They contend friendships are superficial and perhaps false. Some are. But the *Screen Society* study suggests that friendships created and fostered online are different though not less valuable than the F2F friendships that are considered by traditionalists as superior in all respects. To Screenagers, they're not: they're just different. "I don't consider the friendships I have made online to be any less fulfilling than those from meeting face to face," is how one contributor summarized her attitude towards friends. It serves to remind critics that there is no science of friendship, or a formula for making them, and no absolute standard for evaluating them.

Friend is one of those words that has become vaguer as the internet has grown in influence: it used to mean a person with whom we have a bond of mutual affection. So a stranger we have never met could hardly qualify. Now, it's possible to have ties with people we've never met—and call friends. Remember: "There is intimacy and rapport."

Discovery

Question: Why are you always wasting your time glued to that computer? If you're not staring at it, you're hunched over your phone. And, even when you sit down to eat, you have your iPad or Kindle, or whatever it is, next to your plate.

Answer (verbatim from a participant): I discover many of the things that please me by using the internet: I read a book that mentions another book, so I look it up. I hear about a holiday destination, so I search for routes and hotels there. Someone gives me a hot-pot cooking pot I have no idea how to use so I look up recipes on line and discover there is a hot-pot community sharing ideas. I want to know what exhibitions are on in the museums in Edinburgh when I visit so I look them all up. I use it as a source of information mostly. I also submit all of my work electronically. This would be impractical without the internet. OK?

In digital reality, as in other parts of life, there is a frustrating mixture of good and bad, praiseworthy and damnable. Some Screenagers will appear exemplary apostles of the information age, using the net to access information, then segue to more information, then transition to other sites secreting yet more information … and so on. "This is what the net was meant for!" its supporters cry. Snag is: they're not the only ones out there. It's as if there is a price to pay and it comes in the form of users who take advantage of the cyber free-for-all to access not just maps and recipes, but other less savoury data.

Even so, there is a danger in exaggerating the risks. In 2016, Laura De Fazio reiterated the concern of many: "Young users are, in fact, the most vulnerable inhabitants of the cyber world because of their age, their lack of digital skills, and their lack of awareness concerning the opportunities and risks involved with using technology" (p. 212). Lack of digital skills? Lack of awareness? De Fazio seems to underestimate the sophistication of Screenagers' literacy and perhaps their grasp of the potential menace. Young users are very able to access, read and react to information on, for example, cyberbullying, child abuse, grooming, pornography, revenge porn, body shaming and so on. We will return to this theme later, though it's important to present the view that Screenagers are more able than many assume at defending themselves against predators. Perversely, cyber predators are sometimes Screenagers themselves and, as such, have a command of comparable levels of literacy. Both innocent and diabolical have a presence.

The common denominator, like it or not, is that they all devour information. It could be useful, edifying and enlightening or pernicious, noxious and frightening. 12.5% describe their primary online activity as doing research, with 15% reading news. 15.5% are preoccupied with exchanging email and 13.5% prioritize social media. 12.2% ranked sports news as their number one use. These are all, in different ways, information-gathering pursuits and indicate the zeal for learning. We're using learning in a broader-than-usual sense, of course: the acquisition of knowledge or experience. (11.4% used their devices principally for watching videos, whether from YouTube or streaming services. Shopping was top of the list for 11.3%, with gaming 4.5%.)

"It's what the internet is for," one of our participants reminded us. "The availability of resources and ability to learn new skills and being able to connect with like-minded people around the world make the internet a useful and essential part of the modern world." Did the participant use the word "essential" advisedly? We think so. Screenagers take access to information for granted to the point where they see it as crucial and indispensable to today. Later in the book, we will explore whether it is actually a requisite; in other words, could they live without it? The point to make at this stage is simply that none would like to, many would hate to and that all would find it grossly inconvenient.

We could probably revert to more primitive forms of gaming. Shopping would become manual again. And we'd no doubt get back into the old habit of actually going to the cinema to watch films. But how about the living encyclopaedia of knowledge that is literally at our fingertips. Screenagers have grown and matured in a world where information is plentiful and available. Facts, dates, images, even full texts of books offer themselves to anyone who has a portal.

Readers will assume we're overstating our case. Every time we look around and see young and old looking fixedly at their screens, they surely can't be learning. It depends on how we conceive of learning. As we stated previously, we're working with a broad definition that any acquisition of knowledge, constructive or destructive, counts. In some way, it enables us to become competent or proficient in something. Every time users engage with a screen, they are picking up and digesting data, no matter how glib and limited in its usefulness that data might be. We're not claiming the material Screenagers access is weightier than it actually is. But the process of searching for and assimilating information is germane to their experience and it's impossible to find another generation that has ever enjoyed discovering information so much.

The spirit of the Screen Age is information gathering. Much of that information is worthless; some of it is valuable. But it all qualifies as knowledge and this is arguably the single most important defining characteristic of Screenagers: the propensity to learn. Exactly what they learn is open to question; but they do have a passion. And every time you look at someone peering at a screen, you can remind yourself that they're giving vent to that passion. Many will argue that actual experience of the world, physical encounters and human relationships are more effective ways of learning. Perhaps so. But we fill our heads through various means nowadays.

There's a kind of coda to the story described at the start of this chapter. In 2016 Hallam Teaching School in Sheffield, England, conducted an experiment on 2000 13 to 15-year-old school pupils in which they taught in 12-minute sessions, which were repeated twice and broken up with other, unrelated activities such as juggling or model making. Orthodox knowledge suggests that distractions would work against effective learning. In fact, the opposite happened and the researchers found evidence to support the view that information can be more easily remembered if taught in "spaces" of contrasting activities. So, our student who answered the question on the Uses and Gratifications approach to media consumption was probably doing something similar, though, in his case, occupying two or more spaces simultaneously. For more on this, check Richard Vaughan's article "Juggling between lessons boosts learning", in the *i* newspaper, March 31, 2017.

References

Bakó, R. K. (2016). Digital transition: Children in a multimodal world. *Social Analysis, 6*(1), 145–154.

Berger, P. L., & Luckmann, T. (1967). *The Social Construction of Reality: A Treatise in the Sociology of Knowledge*. New York: Anchor Books.

De Fazio, L. (2016). Young people and cyberspace: Introduction. *European Journal of Criminal Policy and Research, 22,* 211–217.

Elwell, J. S. (2014). The transmediated self: Life between the digital and the analog. *Convergence: The International Journal of Research into New Media Technologies, 20*(2), 233–249.

Flores, A., & James, C. (2012). Morality and ethnics behind the screen: Young people's perspectives on digital life. *New Media and Society, 15*(6), 834–852.

Graves, M. F., Juel, C., & Graves, B. B. (2011). *Teaching Reading in the 21st Century: Motivating All Learners* (5th ed.). London: Pearson.

MacNeil, R. (1983). Is television shortening our attention span? *New York University Education Quarterly, 14*(2), 2–5.

Postman, N. (1985). *Amusing Ourselves to Death: Public Discourse in the Age of Show Business*. London: Penguin Books.

Vaughan, R. (2017, March 31). Juggling between lessons boosts learning. *i*. Available at: http://bit.ly/2n1CCXn. Accessed March 2017.

Zilberstein, K. (2015). Technology, relationships and culture: Clinical and theoretical implications. *Clinical Social Work Journal, 43*(2), 151–158.

4

Addiction

Under Control

"The use of mobile telephones releases the same instant gratification feeling hormones as smoking or drinking," suggested one of the *Screen Society* participants. Biochemistry doesn't back up his point, though the underlying idea that phone users, smokers and drinkers all crave adrenaline rushes (immediate gratification) is an interesting one and one that's led people to wonder if we can actually become addicted to screens.

The British term for enthusiastic viewers of television in the 1960s was *telly addicts*. It was a light-hearted expression that was never intended to suggest that devotees of the relatively new medium were dependent on what was, after all, a pleasurable activity. They might have come to depend on the goggle-box (another term from the 1960s) for their evening's entertainment; but this didn't implicate them in a debilitating condition comparable with drug addiction. This was the only kind of genuine addiction known at the time and it was thought to be a strong, irresistible inclination or proclivity to use recreational drugs, particularly heroin.

© The Author(s) 2018

E. Cashmore et al., *Screen Society*, https://doi.org/10.1007/978-3-319-68164-1_4

So, the concept of people developing a habit resembling drug addiction was facile: it was never supposed to convey the true complexities of a pastime that was to become ever more important—to many people—in subsequent decades. To call someone a telly addict meant they spent an unusually long time watching tv. It was intended as a joke, not an attribution of some biochemical maladjustment.

Today, addiction has a much softer nimbus: there are addictions to shopping, sex, eating certain kinds of foods (especially chocolate) and a potentially limitless number of other practices that have been designated addictive, including mainstays like alcohol and tobacco consumption—which were once habits rather than addictions. (We'll come to gambling and gaming compulsions in Chapter 9.)

In our new world, some social habits have been transmuted—by which we mean changed in nature, or form—into medical conditions. Using a model of addiction supplied by substances like drugs, many human practices that involve no drugs, foods, or even other living creatures, have been turned into cravings, compulsions, weaknesses, types of enslavement; in short, addictions.

We rarely question this blurring of activities. One model of addiction refers to altered neurobiology when changes have been initiated by repeated use of a substance. Infants born to mothers who are habitual users of opioids, such as heroin, or non-opioids but still addictive drugs like cocaine and who have used the drugs during pregnancy are sometimes born with dependencies. Innocents that they are, they are born with a physical dependency and need to be weaned off drugs (that is, detoxified) in order to survive. Some children are born with biological or physiological structures that resemble those of their mothers. Thomas R. Kosten and Tony P. George, both medical practitioners, write of what they call "brain abnormalities" resulting from long-term use of heroin, oxycodone and other morphine-derived drugs. But they distinguish these from dependencies, which may also result in abnormalities, though ones that can usually be resolved "within days or weeks" after detoxification. Kosten and George's article "The Neurobiology of Opioid Dependence: Implications for treatment" was published in 2002 and discusses changes in brain processes; they distinguish between "drug liking, tolerance, dependence, and addiction." The more usual assumption is that everything can be grouped as addiction.

So, while addiction, when used thoughtfully (as it is by Kosten and George), can be a valuable concept, its spread has led us towards labelling any repeated behaviour that appears to some people to be problematic as addictive. Anything, literally *anything* can induce a dependency. We've mentioned eating chocolate, watching television and other habits that are intrinsically rewarding and promote repetition. Sex is probably a different category of behaviour, but let's leave it in for the moment. It barely needs stating that there is pleasure taken from these activities, especially when the person has choice. Dependence doesn't involve choice, of course: it describes a state of relying on or being controlled by something (or someone). It doesn't imply changes in brain processes, as addiction does. But it does insinuate a loss of conscious control, perhaps even, in extreme cases, of freewill. When people say they have "demons," they're aware of their own compulsions yet feel they have no choice but to yield to them.

A problem with many useful words that are employed to describe something complicated but with simplicity and clarity is that they're prone to usage creep. And we stand accused: we're extending *mission creep*, which refers to the gradual shift in objectives during a military campaign. Words shift in meaning as they are used in popular discourse. Addiction was once useful (probably) in capturing the manner in which people repetitiously engaged in activities that had deleterious consequences. Their seeming lack of control over their behaviour made it appear that they were literally addicted and had undergone biochemical, perhaps neurological changes that were permanent, or at least semi-permanent. In other words, it was as if their brains had altered: it didn't literally mean they had altered.

The word addiction was useful as a metaphor, a figure of speech, not an exact description. Over the years, we've tended to use addiction as one type of condition that covers all sorts of cases that resemble each other behaviourally. Engaging in sex, shopping, gambling, eating certain types of food and so on brings rewards to certain people. They no doubt *like* to pursue the rewards. They can build up a *tolerance* of them to the point where they need to engage more frequently to experience the rewards. They might even incur *dependence*, meaning they feel compelled to pursue the behaviour and feel helpless to resist.

So their behaviour resembles that of a person who is properly addicted. But the state of dependence is not literally the same: there is no assumption that the person has neurochemical abnormalities, nor that he or she has lost volition, that is the power of using one's own will. To use a term of today, the person still has *agency*, meaning the person can act in a way to produce results he or she wishes. They just don't exercise it in a way that pleases observers; and sometimes themselves, as this *Screen Society* participant remarked: "I often find myself scrolling through my social feeds with no real meaningfulness. If I were to track my time on task, I think I'd be shocked and slightly worried." But another participant issued a veiled and witty caution: "The art of conversation is changing." Note: not "dying" but just "changing". Changing so dramatically and so quickly perhaps that its new structure has become invisible to onlookers, who simply look from afar and conclude: junkies.

The Dark Side

"Mobile phone addiction has increased in recent years." The statement is not ours but that of Li Chen and a research team of psychologists from New York and Wenzhou, China. The team draws on a body of research on mobile phone addiction, its antecedents, its impact on users' academic performance, leisure activities, behavioural problems, health issues and interpersonal life. The focus of the research team is "negative emotions." While the research in itself may be interesting, its uncritical acceptance of mobile phone addiction, which they abbreviate to MPA (as if initialism somehow confers validity), surely needs examination.

A ten-strong research team from Taiwan's Institute of Brain Science, in 2015, concluded: "Smartphone addiction consists of four components, tolerance, withdrawal, compulsive symptoms, and functional impairment." These were offered as "the first proposed diagnostic criteria for smartphone addiction." Examples of functional impairment included: "Smartphone use in situations in which it is physically hazardous (e.g., smartphone use during driving)." Tolerance was identified when "a markedly increase in the duration of smartphone use is needed to achieve satisfaction." Critics might wonder at what point does the kind of everyday user most of us see on our way to work or study or standing in a café or a bar, or in practically any space, public or private, become an addict? Most of the criteria seem widely applicable. But, let's persist with the research.

The premise underlying the concept MPA is that people, especially young people, use their phones for entertainment or, as a related piece of research puts it: "to relieve stress." "Such use can yield immediate gratification," declare Yu-Kang Lee and a research team from the National Sun Yat-sen University in Taiwan in a 2014 article titled 'The dark side of smartphone usage: Psychological traits, compulsive behaviour and technostress.' "But it can also be accompanied by a diminished sense of volitional control and introduce persistent activity" (p. 373).

Like much research in this area, there is evidence of a "diminished sense of volitional control" rather than an actual loss of volition. In other words, a perception; as someone might experience a sense of well-being and happiness, or sense that Tottenham Hotspur will win the Premier League next year. It's an intuitive or acquired awareness.

If the research on addiction is to be accepted, a debilitating compulsion has been taking over the world and has swept up everyone, young and old. It started with drugs, then alcohol and other toxic substances, then food that was supposed to sustain us but which turned out to be junk. After substances, technology kicked in and practices that were once thought recreational pastimes became addictive: television was merely a prelude. Then other long-established habits were upgraded from mere customs to symptoms of bedevilling fixations: yesterday's worker who liked a flutter on the horses with a part of his wages became today's gambling addict with a crippling weakness for squandering money. Have our brains been hijacked either chemically or by some other force?

Mobile phone addiction is but one manifestation of the force. "It's not just our phones we're obsessed with: it's all tech," writes Leslie Goldman in his 2016 article "The trouble with too much tech." It's computers too. Goldman doesn't include tv in what he calls "screen addiction," but probably only because it isn't part of the study that is the nucleus of his argument. That's a survey of 2300 parents of whom 27% were logging a total of eleven hours per day in front of their screens. This included two hours a day doing something on their phones and almost four hours daily on the computer. This equates to screen addiction, according to Goldman. The argument finds favour with an English woman in her fifties from our study, who believes parents actually contribute to the addiction:

> I am concerned that young people age 0-30 will have no idea on how to learn how to develop social skills. I despair at seeing families out for meals and tiny children are handed devices to keep them quiet rather than involve them in the social reality of actually spending quality time interacting as a family.

"Slash screen time in favour of face time," advises Goldman, arguing that parents in particular should "enjoy more meaningful interactions" with their children. This would bolster their self-confidence, language and social skills, as well as "reducing their chances of getting hurt" (p. 72). We presume he means the children rather than the parents, though it might apply to both.

Goldman's argument is, like much prescriptive writing on screens, based on the same addiction model. "We check Facebook while our kids play at the park, take calls during family dinner and – worst of all – chat or text with precious cargo in the backseat" (p. 72). Readers may already have spotted the lacuna in his argument: the kids who are apparently being neglected are doing the same things while playing at the park, sitting at the dinner table or riding in the back of the car.

Of course, this is no defence. Goldman and his allies would simply expand their argument and conclude we've all been zombified by our screens and sacrificed any hope of "meaningful interactions." It may sound bold conjecture but those meaningful interactions are actually taking place. What Goldman and indeed all other critics who follow his directions miss is the detail: it looks like their children are staring unblinkingly into a void, but, on closer inspection, they're socializing.

Socializing is one of those words that has gone out of fashion, but should be brought back—with some revisions. It means, as readers will know, to mix socially with others. Social scientists have their own, more precise definition and use the term to capture the process through which we learn to behave and think in a way that's acceptable to society. Both definitions work in this context. When people appear to be staring inertly at their screens, they are socializing. Their feet may be on the ground, but they are socializing at distance, often with people they have never met face-to-face and perhaps will never share the same land mass. But, thanks to the magical oddity of the internet, they can meet, inter-act, converse, fraternize, consort, mix, mingle and hang with people they find interesting and communicative. What's not social about this?

How does this square with the conclusion of nearly 30% of *Screen Society* participants who believe internet technology and smartphones are addictive? Only 5.4% rejected the idea and the majority (65.2%) opted for the possibility that they can be addictive "if you let them control your lives." Those who thought they definitely were addic-tive believed there was a slight, though significant difference between screen addiction and other forms. As argued by a young woman from Scotland:

> Anything can become addictive if we gain pleasure from it and find it enjoyable. However, with the internet, because it is so ingrained in most of our lives, it is almost impossible to stop using it. Even if its use is becoming problematic we are unable to reduce our internet use as we use it for work, school and social interactions.

This is a clever way of understanding the seeming parallelism between substance addictions which entail a physical dependence and screen addiction: "It is almost impossible to stop using." Slightly less fatalis-tic responses stress our ability to limit our own use simply by exercis-ing self-control: "Had an awful experience with a MMOG that I got hooked on and probably spent upwards of eight hours a day on for a while, finally kicked it. Never again!" (MMOG, or massively multi-player online game, is an online game capable of supporting many play-ers at the same time.) Proponents of the chemical hook-type theories of addiction would find plenty of support from a sizeable number of participants in the study. "I don't think it's a great logical leap to suggest that dopamine is released when we get a notification for a tweet/text/email/whatsapp. That makes it addictive quickly," remarked a young man in his early twenties from Glasgow.

Some participants accept screens have addictive properties, but these don't become apparent until the user is deprived: "People can't be without their smartphones for any length of time before they get anxiety." This is a point taken up by academic researchers, such as Mohamed Salehan and Arash Negahban, of the University of North Texas in 2013. But anxiety is another word with elastic qualities: it can mean a nervous disorder marked by excessive uneasiness and apprehension even in the absence of rational reasons (sometimes called "panic attacks") or it can mean feelings of uncertainty or worry about something in particular, such as losing one's phone or tablet.

One of our participants who considered screen users more balanced than popular images depicted them, said: "Time creates more opportunity - if you have spare time you use the net more." This seems to endow users with more basic good sense than usual. But it's also a clue as to why so many researchers and commentators assume users have surrendered control. The weird enchantment of screens seems a sinister captivation to outsiders; to users themselves, it's a sociable captivation. "Each individual interaction with a smartphone seems quick and simple, but is rewarding - a sure fire recipe for addiction!" joked one participant in his late twenties from London.

One Screenager went further and sneered at the very suggestion that screens are addictive:

> If information is power and power is addictive, then having access to the breadth of information available via the internet through the use of devices is addictive. Personally, I find it difficult to not look at the internet for breaking news stories.

Considering the overwhelming number of users who use their devices primarily for accessing news, this seems a reasonable rebuttal: if familiarizing oneself with up-to-date information about the world is addictive, is this so unnatural, or is it, as this participant detected the product of human curiosity?

I think we all have a curiosity to find out what's going on (in our own worlds) and that is what can be addictive. People used to insist on everything stopping while the news was on the tv or radio. Now, in a similar way, I do check various sites before I relax for the evening without a screen. I use spoken word podcasts on an iPod to open the world up to me and this is an excellent improvement on having to listen to whatever radio show was on at the time you wanted to listen.

The Opposite of Addiction

The writer Johann Hari recounts the story of an intriguing piece of research and its coda. "The experiment is simple," writes Hari. "Put a rat in a cage, alone, with two water bottles. One is just water. The other is water laced with heroin or cocaine." Guess what. The rat kept returning to the water with the drugs. (Nowadays, the rat would probably turn its nose up to tap water anyway, preferring Evian or some other branded H_2O.) The organization, Partnership for a Drug-Free America featured the experiment in its advertising: "9 out of 10 laboratory rats will use it"—the spiked water, that is. They drank and drank the cocaine-infused liquid until dead. "And it can do the same thing to you." The ad intentionally highlighted how, if rats can become addicted so easily, so could humans.

Any self-respecting researcher immediately hears an alarm bell and a warning sign lit up with "false analogy." There is a well-known trap called the logical fallacy and all researchers guard against applying the results from one situation to another, especially if the situations are significantly different. And rats are, by common (though not unanimous) agreement, significantly different to human beings (Readers who have had their appetites whetted by this might find Espen A. Sjoberg's 2017 article "Logical fallacies in animal model research" instructive).

This was by no means the only weakness in this study. What about if we took the rats out of the sterile austerity of a lab and gave them an altogether plusher, more comfortable environment where they would have all manner of coloured balls and other toys and the best rat food money can buy? Bruce Alexander, a psychology professor from

Vancouver, created a palatial home for his rats and then supplied the two water sources. The water was still from the tap (it was the late 1970s) but the rats still tried both. But, whereas the rats in the early experiment kept returning to the dope until it killed them, Alexander's rats, which were living high on the hog remember, preferred the unadulterated stuff.

Hari's purpose in relating this is to question orthodox accounts of drug addiction: since the end of World War II, addiction has developed into a major narrative stretching across several decades. The popular account holds that, following the war, a pleasure seeking, self-centred and narcissistic generation sought stimulation from anywhere and anything they could. The effects of drugs were rewarding in the short term and induced more people to use them. The long-term effects were more damaging. But addicted users found it hard, if not impossible, to wean themselves off whatever narcotic they have become accustomed to using. Hari's counter to this is that the Vietnam War was the equivalent of the lab cage the first rats were locked in. Returning from the war having become hooked on drugs while in military service, the US soldiers brought their habits home with them and created a demand for drugs. The global cultural diffusion of drugs started, on this account, around 1973, when the Vietnam War was over.

We aren't going to argue whether Hari's general theory of the spread of drug use and the method of addiction should be accepted in their entirety. He does, however, extend his argument with the thought-provoking idea that addiction can be understood as a form of bonding. A perverse form, we might add. "A heroin addict has bonded with heroin," writes Hari, presumably licencing readers to substitute the drug with practically anything. The sex addict is bonded to sex, the alcoholic to alcohol, the shopper to shopping etc. Bonding is a straightforward word but one loaded with ambiguity. It describes a feeling or perhaps an invisible force that brings and keeps people together; that feeling or force may be shared emotions or common interests. Whatever it is, it yields an affiliation, or attachment that keeps people close and, for the most part, contented, if not pleased.

The reason Hari's hypothetical addict bonded with heroin is "because she couldn't bond as fully with anything else." This might sound a bit sweet, simplified and a tad too neat. The implication is that, if we all bonded more satisfactorily, there would be no addictions in the world: we would all forge fellowship, coalition or close associations with other humans and have no need or desire to become chained to substances or destructive habits.

This chimes with the view of one participant who, when asked whether screen devices are addictive replied: "No more than there is a human need for company and companionship. This just provides an alternative." Hari's conclusion is bracing: "The opposite of addiction is not sobriety. It is human connection."

This has interesting implications, even if we don't accept that human connection is exactly the opposite, that is totally the reverse, of being physically and mentally dependent—autonomy, independence or self-determination would seem to fit better. But semantics apart, Hari's point is a strong one: it encourages us to recognize that, instead of being an addiction, screen using is actually an effective alternative; it allows us to build the kind of bonds that prevent, discourage and deter addiction. Connectivity takes many forms, but they all boil down to the state of being connected, or, to be strictly accurate, interconnected (*inter* meaning between, suggesting reciprocity and interactivity).

Social connectivity, or, to use the old-fashioned term we favour, sociability, is one of the motivations that draws Screenagers to a world that is ordered, coloured and populated by people with whom they share interests and tastes. Another, as we mentioned previously is the hectic search for information. As one participant put it: "Fast pace of change excites people desperate to find out about latest updates." Knowledge never stands still, of course. The desire to keep up implicates users in a way that's often mistaken for pre-cognitive inertia. It's anything but. As one respondent sarcastically pointed out: "I certainly find it [screen time] addictive, and wonder what I filled the time with before; perhaps it's down to the human need for knowledge (Not always meaningful information)."

"Human beings are bonding animals," propounds Hari. "We need to connect and love." The sentiments will probably be too idealistic for many, but there's value in the aspiration: if we could create an environment in which human beings can satisfy our yearning to relate to each other, the temptation to bond with alternative, enfeebling things and practices would disappear. Addicts of whatever kind rarely complete a rewarding day's work, return home to comforting surroundings and relax with fulfilling recreational activities. We don't have to accept all of Hari's argument to see its relevance to the present topic. Screenagers are, on one account, addicted or at least approaching addiction; on their own view, they immerse themselves in a social environment in which they can position themselves as they wish and interact with other like-minded enthusiasts. This is strictly not a passive exercise. (The rat studies Hari references are the subject of a YouTube video here: www. youtube.com/watch?v=7kS72J5Nlm8.)

Habits and Choice

The only reason to make the distinction between a habit and an addiction, wrote the Hungarian-American psychiatrist Thomas Szasz, "is to persecute somebody." Szasz (1920–2012) never tried to conceal his skepticism about the medical profession. He believed all human beings are free agents, responsible for their actions and, by implication, for their consequences. He denounced any incursions on civil liberties in the name of psychiatry.

Szasz's influential 1961 book *The Myth of Mental Illness* advanced the controversial argument that mental health and mental illness were pseudo-scientific, pseudo-medical terms, and that that illness, in the modern, medical sense, applies only to physical bodies, not to minds; we forget this when we talk of mental disorders as illnesses or diseases. Szasz insisted that these are metaphors and not literal descriptions. What Szasz would have thought about so-called "internet and computer addiction" we can only imagine. But, if presented with the evidence from our study, we suspect he'd be persuaded that screen addiction is a pseudo-condition used typically by people who don't use screens much, or don't understand them, or don't want to understand them, or willfully try not to understand them because they make a decent living out of critiquing what is, to them, unintelligible behaviour.

Screenagers habitually look at and use their screens. Repeat: habitually. Reading, swiping, tapping, talking and listening are practices, patterns, tendencies or gestures. They might appear as if they're responses to cravings, fixations, compulsions, obsessions or addictions. Until, that is, their meaning is discovered. Instead of suffocating reason, interacting online might do the opposite—oxygenate it, perhaps.

It's naïve to be surprised. By anything. Nothing should really amaze, astonish or stupefy a twenty-first century population reared on DNA sequencing, genetically modified plants, biometric passports, magnetic resonance imaging, photovoltaic solar energy, stents and ATMs.

Seeing several generations of people staring at screens shouldn't surprise us into a hurried and thoughtless conclusion. Screenagers are not addicted. In fact, they are, if we accept Hari (even in part) participating in an activity that is, in many respects, contrary to the passivity and fatalism that addiction usually connotes. None of this means that addiction is unthinkable for many screen users; for some, their screens do sometimes loom like those shiny pendants stage hypnotists swing in front of their would-be subjects. "As with anything you need self-control to stop yourself from becoming addictive," said one respondent, who accepted that her engagement with her screens could become a preoccupation. Another participant is more resolute: "Nothing is addictive unless you allow it to be." And another who thinks, if Kim Kardashian carries a Prada bag, the rest of the planet doesn't have to rush out and buy one, simply says: "Not all people are sheep! Baa!"

A central part of this book's argument is that critics and commentators on screen, or perhaps the screen culture that has evolved in recent years, is a glimpse of hell on earth. Our streets are inhabited by what seem to be reanimated corpses turned into creatures capable of limited movements but not of sentient life. We admit: it does look harrowing; there is a *Dawn of the Dead*-like quality to our urban landscape. Our task in this book is to clarify what is going on. Almost every participant in our study agreed that screens do carry the potential to "suck you in," as one contributor put it. No one denies the mesmeric appeal of screens. Equally, only a tiny number believed users could become helpless dupes. Willpower, as many participants observed, was an unfashionable and underutilized capability that we tend to neglect when discussing the multiplying number of addictions. "We still have the ability to make the choice - it's *not* exercising the choice that gives the appearance of addiction," said a New Zealand woman.

Type "computer addiction" or "internet addiction" or "screen addiction" into your search engine and, while you won't (or shouldn't) be surprised by the number of hits, you will realize just how seriously these concocted maladies are now taken, particularly by medical practitioners and university scholars. Crass terms like "digital heroin" and "electronic cocaine" are now used (check Dr. Nicholas Kardaras in the *New York Post*, August 27, 2016). It probably won't be long before the American Psychiatric Association's *Diagnostic and Statistical Manual of Mental Disorders* (*DSM*), which is widely considered the authoritative reference source on conditions that disrupt mental functions, integrates screen-related afflictions into its text.

Our choice is to look beyond the screen itself, to grasp the conversational exchanges that feed the appetite for information and the social bonding that Hari believes is an antidote to addiction.

References

Chen, L., Yan, Z., Tang, W., Yang, F., Xie, X., & He, J. (2016). Mobile phone addiction levels and negative emotions among Chinese young adults: The mediating role of interpersonal problems. *Computers in Human Behavior, 55* (Part B), 856–866.

Goldman, L. (2016, August/September). The trouble with too much tech. *Scholastic Parent and Child, 22*, 72–76.

Hari, J. (2015, January 20). The likely cause of addiction has been discovered, and it is not what you think. *Huffington Post*.

Kardaras, N. (2016, December 17). Kids turn violent as parents battle 'digital heroin' addiction. *New York Post*. Available at: http://nyp.st/2n9IiKS. Accessed March 2017.

Kosten, T. R., & George, T. P. (2002). The neurobiology of opioid dependence: Implications for treatment. *Addiction Science & Clinical Practice, 1*(1), 13–20.

Lee, Y. K., Chang, C. T., Lin, Y., & Cheng, Z. H. (2014). The darkside of smart phone usage: Psychological traits, compulsive behaviour and technostress. *Computers in Human Behavior, 31*, 373–383.

Lin, Y. H., Lin, Y. C., Lee, Y. H., Lin, P. H., Lin, S. H., Chang, L. R., et al. (2015). Time distortion associated with smartphone addiction: Identifying phone addiction via a mobile application (App). *Journal of Psychiatric Research, 65*, 139–145.

Salehan, M., & Negahban, A. (2013). Social networking on smartphones: When mobile phones become addictive. *Computers in Human Behavior, 29*, 2632–2639.

Sjoberg, E. A. (2017). Logical fallacies in animal model research. *Behavioral and Brain Functions, 13*(3), 1–13.

Szasz, T. S. (1961). *The Myth of Mental Illness: Foundations of a Theory of Personal Conduct*. Cambridge: Harper & Row.

5

Politics

Politicians' Thoughts

"Believe only half of what you see and nothing that you hear." It's a quote from Edgar Allan Poe's *The System of Doctor Tarr and Professor Fether* and was endorsed by one of the *Screen Society* participants, a middle-aged man from England, who repeated the quotation then added, "*a bit like politicians then.*" Those who took Poe seriously would have believed nothing that passed politicians' lips before the late 1950s. But television made it possible for them to believe 50%.

Since the famous televised John F. Kennedy and Richard M. Nixon American presidential debates of 1960, there's been little doubt that the screened image can be more persuasive than the spoken or written word. In this instance, it seemed to confirm the old saying, "The camera never lies." Or did it? Nixon held his own in the tv debates and the majority of those who listened to them on radio believed he came out on top. But, on tv, his ghostly pallor and jowly cheeks made him appear a less attractive candidate than his fresh-faced opponent who emerged triumphant in the election. At the time of the Kennedy–Nixon debates, the printed medium was considered the most credible source of news.

© The Author(s) 2018
E. Cashmore et al., *Screen Society*, https://doi.org/10.1007/978-3-319-68164-1_5

Despite its domestic growth over the previous decade, television was still something of a novelty and lacked the gravitas of newspapers and journals. We've since grown evermore reliant on television for our political information, as we have for every other kind of information.

Kennedy was the first modern politician to realize the potential of television in politics. Gil Troy argued in 1998 that the supposed polarity between commander-in-chief and celebrity-in-chief is a false one. Presidents are policy-makers, he argued; but, after JFK, they had to be "popular figures" too. Troy was writing at the end of the twentieth century, remember.

Technology was making its impact on politics, as it was on every other sphere of society. Television in particular was familiarizing us with politicians in a way that, for want of a better word, humanized them. We could see and hear them and watch their body movements, trying to detect any telltale signs. But we didn't know what was going on inside their heads. Now, we think we do. If we want to check what thoughts are circulating in the minds of our leaders or prospective leaders, we just read their personal twitter accounts. "It gives people a good insight into politicians' thoughts … Allows politicians to be seen as familiar and relatable and interact with the public," explained a young female participant in her early twenties in the *Screen Society* research.

"Politics has been transformed by communication, soundbites and media friendly politicians. Each new technology, be it radio, tv, websites, social media, means the boundaries change; and politicians keep apace with it or lose out in elections to others who have," a middle-aged Belfast man mused. In the 1920s, radio separated political voices from their physical form. "Politicians, used to bellowing at fairgrounds and train depots, found themselves talking to families in their homes," reflects Nicholas Carr, of *Politico Magazine*. "The blustery rhetoric that stirred big, partisan crowds came off as shrill and off-putting when piped into a living room or a kitchen." Winston Churchill mastered the radio as perfectly as any politician, particularly during World War II, when his voice galvanized listeners. His rousing "We shall fight on the beaches" speech on BBC, June 4, 1940, is still arguably the most powerful political broadcast in history.

It might be a far cry from Churchill's evocative use of a medium and stands no chance of living as long in the collective memory, but Donald Trump's employment of media was memorable nonetheless. Even if only as an example of how methods of communication are as important as content. On August 26, 2016, when the American presidential rivalry between Trump and Hillary Rodham Clinton was approaching its climax, Nykea Aldridge, an African American mother, was killed by a stray bullet during a gang shootout in Chicago, a city Trump had repeatedly identified as exemplifying the USA's troubles. Aldridge was a cousin of NBA player Dwayne Wade and had been pushing her son in a pram. Within 24 hours of the shooting, Trump tweeted: "Dwyane [sic] Wade's cousin was just shot and killed walking her baby in Chicago. Just what I have been saying. African-Americans will VOTE TRUMP!" Later, he tweeted again, correcting his misspelling of Dwayne. Still later, he conveyed his "thoughts and prayers," again via twitter. The retweets went into several thousand.

Critics lambasted Trump for his naked opportunism and tasteless attempt to make political capital from personal tragedy. But Trump was unrelenting and, later, reflected on twitter: "It is so nice that the shackles have been taken off," an allusion to the freedom and directness facilitated by social media. Other politicians had used twitter effectively before Trump, but the Republican celebrity—he had previously not held political office and was known principally for his reality tv show—relished his ability to communicate with the electorate without the orthodox filters operated by media advisers. "Information is no longer the province of a few news outlets," concluded one of our participants who favoured the changes to political discourse ushered in by social media.

There's substance to support this: we've witnessed politicians sharing Snapchat stories, tweets, and YouTube videos as well as live streaming on Facebook and Periscope. Trump himself was the first politician optimized for the Google News algorithm. There is no shortage of information from politicians. The quality or value of information is open to question. Most politicians issue unexciting updates, while others deliberately provoke with insults or incendiary material. At the time of writing, Trump's description of Clinton aide Human Abedin as "the wife of perv sleazebag" is probably the gold standard.

We never know for sure whether politicians disclose their true selves (whatever they may be) on the internet. In 2016, Trump probably came close: he outraged and offended voters time and again, refusing to moderate his language or tone. Other politicians pay only lip service to social media, sending out anodyne messages that seem to provide little if any insight into how they really think about issues. Perhaps, in years to come, people will read Trump's tweets in the same epoch-defining way they read Lincoln's diaries or Kennedy's journals. We are not being facetious: America's National Archives and Records Administration, in 2017, told the White House to keep copies of each of President Trump's tweets, even those he deleted or amended, according to journalist Stephen Braun. (Question: doesn't twitter archive tweets automatically?)

The screen offered politicians what was, in the 1960s, a unique way of influencing voters—and that's basically what any politician seeks to do: affect the behaviour of people who can vote for or against him or her. In the days of television, politicians would give interviews or offer political statements directly to cameras, or even participate in studio debates, much as they do today. Television disrupted the previous relationship between politicos and electorate, which used to be based on public assemblies and interviews with chosen journalists that were transcribed or interpreted and published in magazines or newspapers. Radio modified this by allowing listeners to hear politicians speak and answer questions. But actually seeing them on screen, watching them squirm, sweat anxiously, fidget uncomfortably or perhaps luxuriate in an environment that suited them perfectly (as JFK did), permitted an evaluation based on more personal qualities. Viewers could actually see their leaders, even if they couldn't read their minds—as our participant believes.

Interactive media must have arrived like a plain wrapper parcel on politicians' doorsteps: what was in it? Valuable gifts from exotic places? Or explosives? As reported in a Haroon Siddique article for the *Guardian* newspaper, former British Prime Minister David Cameron seemed to assume the latter when, in 2009, he answered a radio presenter's question about his views on social media with: "The trouble with twitter [is] the instantness [sic] of it – too many twits might make a twat." Barack Obama, US president between 2009 and 2017, favoured the former: he garnered 86 million+ followers with his vigorously expressive tweets that seemed to maintain an open portal to his state of mind. Obama, more than any other politician, showed how to reach a generation that had, by 2009, started to turn away from traditional news sources and switch to online platforms. Screenagers, as we discovered from the project, don't regard social media as a distinct and unusual medium, one among many: for them it is the go-to source; social media is actually not *social* media; it is just the *media*. So, it's only sensible for politicians to squeeze every advantage they can out of this digital gift (rather than bomb).

British Labour Party leader Jeremy Corbyn, acknowledged this on his Facebook page in May 2016, slyly noting how previous generations had been denied full access to left-wing politicians by establishment media institutions: "When in one week, we can get 1 to 2 million people watching online a message … it is a way of reaching past the censorship of the rightwing media in this country that has so constrained political debate for so long."

It was a redoubtable piece of socialist critique, though Corbyn neglected the extent to which mainstream media and social media share content. If, as Corbyn supposed, "the censorship of the rightwing media" inflected, if not distorted, political reportage, then it's impact would probably be enhanced by its distribution in social media (for more on this, check Jim Edwards article "Jeremy Corbyn's Facebook Strategy is so much more sophisticated than you think", in the Business Insider UK, June 2016). So, there is as much reactionary as radical potential secreted in digital media. There is also a limitless potential for downright untruths, as we will discover later in this chapter.

If television glamourized politicians, the internet deglamourized them, indicating how media technologies have the capacity to transform the manner in which our elected leaders address us—and, by implication, us them. Though, as one participant detected, there is a constant: "Politicians will always lie no matter what medium they are using." (This echoed the views of over 65% of participants.)

Perhaps voters have always held their elected leaders in low esteem. But the suspicion is that media, or at least screen media, have applied more downward pressure. "Different media, same old nonsense," moaned one participant when asked to discern the changes in substance in the transition from traditional to digital. Another one added: "So the internet is going to tell me when a politician is lying? I don't think so." Obviously, there are still plenty of advocates of Poe's words of wisdom.

Opening Eyes

"Today, with the public looking to smartphones for news and entertainment, we seem to be at the start of the third big technological makeover of modern electioneering," writes Nicholas Carr. "This shift is changing the way politicians communicate with voters, altering the tone and content of political speech." One of our participants agreed, qualifying with: "It creates more controversy and increases in propaganda." Certainly online political discourse seems to incite controversy rather than merely inform (though it could be argued that this in itself provides useful information).

The movement away from traditional broad-based party politics towards the formation of exclusive political alliances around particular people or issues had been going before the rise of internet. But portable devices have taken politics on a different tangent. Professor W. Lance Bennett of The University of Washington observed "more diverse mobilizations" among electorates around the world. These are "mobilized around personal lifestyle values to engage with multiple causes such as economic justice (fair trade, inequality, and development policies), environmental protection, and worker and human rights."

Even this sounds outdated. Bennett recorded his observations in 2012, six years after the launch of twitter and four years after Barack Obama had shown how to proselytize online. Proselytize usually means to convert or at least try to convert people from one religion to another, but it can also mean to convert them from inactivity to political engagement, which is exactly what Obama did. His campaign garnered five million supporters on various social platforms and he managed to raise money grow a body of volunteers, all presumably believing they could make a difference. He also galvanized donors and advocates, augmenting his tweets and online messages with a stream of text messages. Obama won the election by nearly 200 electoral and 8.5 million popular votes. In the process, he underlined the role digital media would play in politics.

This sounds as if it is, or at least should be beneficial, especially if we consider the various points made in the preceding portion of this chapter: politicians have been brought down to earth, forced to make themselves accessible to their voters and practically compelled to communicate if not face-to-face, then digitally. But not everyone welcomes this development. One of our contributors reflected on the changes wrought since Obama's first election: "Politics has already been transformed. Twitter has distilled political debate to 140 characters. Attention spans have shortened, and the soundbite is now driving society."

One of this book's arguments is that the sense of alarm occasioned by digital media echoes the alarm heard when television appeared and threatened to traumatize personal and public life. That included politics. The Kennedy–Nixon tv debate was notorious because it seemed almost tantamount to cheating: if a thrillingly handsome man with a certain élan could wipe the floor with an erudite, but unattractive type with pale skin and thinning hair, then maybe visual panache was strong enough to ambush political ideas. And that was just one of the many ways television could change politics for the worse.

Each period of panic or hysteria has its moment, then passes on, leaving society a little different. Today, it is would be unimaginable to have an election in which the candidates refused to communicate with the media. Television has shaped our world just as surely as digital media is shaping it right now. Sometimes the effects are direct. Writing in 2013, John H. Parmelee and Shannon L. Bichard's argument is that twitter became especially potent in the second decade of the twenty-first century when all politicians learned that their followers were not just voters. Journalists also hit that little + Follow button on the right of the screen. Parmelee and Bichard believe politicians use digital media to influence the influencers (i.e. journalists and editors), who can in practice dictate and change agendas. Carr takes this one step further, arguing that Trump in particular showed how provocation beat rational, informative discussion every time when it came to dominating the influencers. In other words, traditional journalists, for long habitu-ated to controlling the menu of political campaigns, found themselves being force-fed nutrition-free fast food (excuse the tasteless metaphor; it is ours, not Carr's).

So, while one of our study participants proposed "politicians can bypass the traditional press and are less likely to be held to account by the press," she may not have noticed how the bypass may be a subter-fuge and that the real, yet undisclosed goal is to connect with traditional media. As we pointed out earlier, the demarcation between traditional and digital news sources is not nearly as clear cut as many assume: online news frequently gathers its news from traditional news organiza-tion. Even provocations designed, we presume, to elicit anger or dispute now qualify as hard news. Scandals, whether precipitated by digital or traditional media, often become political currency. Even media them-selves can become factors. In the French presidential election of 2017, 55% of French voters considered candidate Marine Le Pen had been treated negatively by traditional media. That in itself became a political factor. All of which leaves us wondering whether digital media is a good or bad thing for politics?

Echo Chamber

> Politics has always been something that evolves. Putting politicians on the news, via tv has been incredibly influential in terms of who and what the politician are. Instant communication with the world can be good in that more people are debating with a much wider range of views. On the other hand, it leads to a lot of ill thought out tweets.

It's a balanced view from one our participants and leads us to think in terms of what Edward S. Herman and Noam Chomsky called in the title of their 1988 book *Manufacturing Consent*. It referred to the pro-duction process in which governments, multinational corporations (what we'd call global corporations today) and the mass media (as it was called) assemble a broad consensus that suits the needs of both the market and existing political parties and perpetuates the status quo. The consent of the title refers to the agreement or acquiescence of the public.

Updating the argument to account for vertical integration, in which media companies incorporate separate businesses into a central organization, we could argue that the broadcasting and print media behemoths work hand-in-glove with politics to maintain a cosy alliance. If so, then digital media should, at least in theory, be a force for an authentic democracy. There are various strength versions of this from our participants. A *two-bar* strong interpretation to start: "Politics will not be transformed by it [digital media]. What will happen is: the public will become much more informed about politics on a local and national level, which will change the reasoning behind their votes." *Three-bar*: "More views will allow people to pursue answers to questions they have. Old media will have less sway. Spin doctors will become increasingly redundant." *Four-bar*: "From alt.news and fake news, twitter fights and Facebook posts, people can post opinions and gather support from all over the world, and other opinions can be easily ignored or silenced creating an echo chamber." So plenty of democratizing potential, but ultimately consigned to an "enclosed space." And finally a *five-bar* verdict: "People start to turn away from the mainstream news outlets and seek alternative sources. In an increasingly divided society people believe less and less in the government and news outlets. It is also a good place to connect with people with similar ideas and values which will help people organise and display their displeasure at the status quo."

In 2007, Michael J. Jensen, James N. Danziger, and Alladi Venkatesh used the term "cyber-optimists" to describe people who believe that "the Internet will help mobilize groups of people who otherwise are not politically engaged." They contrasted these with those who "argue that the internet tends to reinforce the same structures that otherwise constitute 'determinants' of political practices offline" (p. 41). While our more recent evidence indicates the continued presence of those who suspect only reinforcement of existing structures, it also indicates a strong belief that online political engagement is already a growing force (we wouldn't call it optimism because that suggests a confidence in rather than just hopefulness about the future).

We'd also caution against the deduction of Jensen, Danziger and Venkatesh: "The digital divide will generally reinforce existing dispari- ties in political participation that tend to favor those with higher SES [socioeconomic status]" (pp. 41–42). When Jensen and his colleagues were writing in 2007, the so-called digital divide was a gulf between those who have ready access to computers and those who didn't. There is no longer such a gulf.

Another indication of how fast the world turns is in the researchers' conclusion: "Our analysis shows that the Internet mediated activities are not simply an extension of offline political practices, but appear to be a distinct, although socially embedded, medium in which politi- cal behavior takes place" (p. 47). Many of our participants saw differ- ently, arguing that one was very much as extension of the other. Online participation is valuable in "enabling rapid dissemination of ideas and connections of like-minded people [thus promoting] rapid protest mobilization," according to a Canadian man in his mid-thirties.

Unlike most books and articles on the net, *Screen Society* is not based on pronouncements or warnings, but on the perspectives of the peo- ple who matter: the users. That currently includes about 90% of the European and American populations and will soon apply to every- one. It's inevitable that politics will change in some substantial way as a result of our screen habits: Screenagers will think and feel differently about politics; they will relate to and interact with politicians in ways we're only just beginning to grasp. So when Carl Miller writes in the 2016 book *The Rise of Digital Policies*: "There is a clear expectation from the digital electorate – an emerging norm in these spaces – that there should be two-way communication between them and politicians," he is barely hinting (p. 32). The question is not whether that expectation will be met—it definitely will—but, with what consequences?

Guardian writer David Runciman believes: "The most striking thing about our politics is how little it has changed." His argument is sim- ple: the surge in populism witnessed in the UK, USA and France in the

2016–2017 period combined with the transformative changes wrought by digital media from the 1990s created the exaggerated impression that the political world had been changed just like every other aspect of society. "The hopes from the dawn of the digital age of a new era of democratic empowerment remain unfulfilled."

One of our participants agreed: "Politics are being reduced to 15 second soundbites and no real political discourse takes place." While a sizeable percent of people who had used social media for politics, reported that they felt more politically engaged as a direct result, most participants believed that benefits would be compromised. Another participant offered a thought, again not dissimilar to Runciman's, but with an inflexion:

> Not only does the internet lead to a favouring of soundbites and easily-digestible, unverifiable statements with little detail to act as news, it offers a platform for activists (for good and ill) of like minds to discuss their beliefs and mobilise with others from around the world who feel the same way.

One of the central premises of this book is that digital media should be analysed in terms of its potential. Like every other piece of technology, it has been conceived and constructed by humans and remains at humans' service. It is ours to use; forget this and we slide towards alarmism. So, when one of our participants states, "I have become much more politically aware using my phone and social media than I would have without," we take it at face value, but remain mindful of the comments of another participant:

> Politics will never be transformed and anyone that believes so needs a slap. It's all fake, they [politicians] all work for big business and corporations, the same people who fund their parties, fund think tanks and the like, which are also held up as independent, when they are highly and ridiculously involved … The internet is never going to create revolution, it can only open the people's eyes who wish them to be opened to the real-life politics and not the fake democracy we have now.

It can only open the people's eyes who wish them to be opened: it's a telling phrase though one tinged with fatalism.

About a third of users are convinced the net's impact will be positive (we approximate the figure because the research didn't attempt to collapse responses into yes/no-type simplifications; our judgement is formed from lengthier expositions).

Myths, Lies and Distortions

Politics in the new, digital world are as much about people as subjects, policies or traditions. This doesn't mean we've all lost our political as well as moral compasses; it means digital times needs digital politics. Excuse the casuistry. Two of our participants expressed much the same idea thus: "People [will] start to turn away from the mainstream news outlets and seek alternative sources." "Old media will have less sway. Spin doctors will become increasingly redundant." Their point, shared by almost everyone, is that people still have a hunger for answers to questions that would previously have been addressed through mainstream television and print media. Now they seek answers through a miscellany of online sources.

For many this is a change that will become more profound in years to come and politics catches up, as they see it, with the changes brought about by technology. But others "fear that the shattering of the old systems is allowing myths, lies and distortions to spread like wildfire through our interconnected screens." This is the view of *Financial Times* journalist Jim Pickard and it's a view that finds favour with many participants. Consider, for instance, this particular response: "People tend to believe what they read on the internet without questioning the source of the information."

The allusion here is to what became known, in 2016, as "fake news." The term was new, but the practice of distorting the truth for political gain is probably as old as politics itself. Octavio famously used disinformation to assist his military victory over Marc Antony in the final war of the Roman Republic, 30–32 BC. Disinformation was and is intended to mislead, especially for the purposes of supporting propaganda. The

British government are past masters, caricaturing Germany as "the Hun" during World War I, 1914–1918, to create a personification of the enemy. Propaganda filtered into popular culture too. For example, a series of Hollywood-made Sherlock Holmes movies in the 1930s and 1940s were very loosely based on Conan Doyle stories updated and infused with anti-Nazi sentiments. One of our participants believed digital media performed much the same function as the Holmes films: "I think it creates more controversy and increase in propaganda." Of course, the post-2016 disinformation was quite different from anything experienced in wartime: it involved relatively small numbers of tech-savvy people exploiting the traffic of digital media interaction by constructing news-based stories, often exaggerated or just made-up, about major political figures. Then just waiting to discover if anyone was smart enough to repudiate them.

The expression, "There are three kinds of lies: lies, damned lies, and statistics" was attributed to a politician, Benjamin Disraeli, who was Britain's Prime Minister, 1868 and 1874–1880. The attribution itself may be spurious: it was from Mark Twain, who probably took it from a speech by another British politician, Leonard Henry Courtney, who probably heard it from Arthur James Balfour, who was Prime Minister, 1902–1905. Whatever the origins of the quote, it's very clever though probably in need of revision: lies, damned lies and what passes for truth might be congruent with modern life.

Digital media dispenses truth and lies in *what seems* to be equal measures; no one knows for sure. It's impossible to discern many patterns from the assorted mishmash of lies, damned lies and so on; but there is one. Politicians and their aides are intent on undermining their political rivals either with hyperbole or the embellishment of truth.

According to James Carson of the *Telegraph* newspaper when writing in 2016, there are five types of fake news: (1) *Intentionally deceptive*: for example, a story that a popular sports or pop star supported a certain politician, or a particularly unpopular celebrity supported an opponent; (2) *Jokes taken at face value*: these might include the famous Obama–Bill Clinton conversation in which the former whispers, "Don't tell anyone but I voted for Trump," and the latter laughs, "So did I"; (3) *Largescale hoaxes*: if someone reported that British businessman Sir

Philip Green, a known Conservative supporter, had made out his will, leaving his £8.5 billion fortune to the Green Party because he thought they named the party after him, it would be nice if implausible hoax (at least, we assume so); (4) *Slanted reporting of real facts*: academic journals are reliable sources of research findings that can be angled to highlight a feature, such as "4 out of 10 Scottish National Party Members are Depressed" from a study of how depression afflicts populations; and, finally (5) *Stories where the "truth" is contentious*: reports from warzones are sometimes based on second or third hand accounts, for instance.

Carson's typology is far from exhaustive and many stories don't fit neatly into a category or possibly straddle several. Some are worthy of laughter, others offer raw material for chatter and still others occasion serious discussion. But for all the discussion they generate, Carson asks the crucial question: do they influence anyone, politically? Before his answer, we'll consider some of our participants' views.

"The internet offers huge scope for diversity but confirmation bias is the biggest problem," suggested this male in his forties from London, referring to the tendency to interpret new evidence as confirmation of existing beliefs or theories. In other words, even strange and perplexing data is assimilated into our frames of reference and supports what we assume to be true. Another participant doubted its impact though for different reasons: "I think politics come from many sources - family, social class, life experience etc. I don't think the internet will influence political opinions."

Yet another (and we repeat): "From alt.news and fake news, twitter fights and Facebook posts. People can post opinions and gather support from all over the world, and other opinions can be easily ignored or silenced creating an echo chamber." This metaphor of an echo chamber effect in which the voices of many are confined in a way that effectively nullifies their influence is a popular one, of course. Carson thinks the proliferation of reports that bend reality to suit specific ideological or practical purposes has affected the way politicians behave and that it will have utility, though perhaps not for obvious reasons. The term fake news, or—to borrow our participant's phrase—alt.news (alt. denotes a version of something that's intended to challenge a traditional version) will be appropriated by politicians "to mean anything they disagree with." This will reduce, if not annihilate, its use, apart from as "a stick to beat the mainstream press with." Its overall political influence will be limited. Voters' scepticism is already at 11, so it's unlikely to be cranked any higher by fake news. We're inured to lies anyway.

People have a completely understandable propensity to believe any kind of information, no matter how outrageous, about politicians. Politicians may imagine we see them as reforming Messiahs, and, perhaps, for a while this is how we actually did see them. As one our participants put it: "In the 1980s and 1990s, the media [television] was hugely influential in projecting 'charisma' candidates in elections." But in the new century, they look more like misfits who probably couldn't hold down a proper job, and found politics a congenial place where their faults could be covered up and their deceptions would never be discovered.

Having opened this chapter with one literary giant, it seems appropriate to close with the words of another. Although he was writing in the 1940s, George Orwell might easily have been commenting on today: "In times of universal deceit, telling the truth becomes a revolutionary act."

References

Bennett, W. L. (2012). The personalization of politics, political identity, social media, and changing patterns of participation. *The Annals of the American Academy of Political and Social Science, 644*(1), 20–38.

Braun, S. (2017, April 4). National archives to White House: Save all Trump tweets. *Chicago Tribune*. Available at: http://trib.in/2o0EXiE. Accessed April 2017.

Carr, N. (2015, September 2). How social media is ruining politics. *Politico Magazine*. Available at: http://politi.co/2o5Megl. Accessed April 2017.

Carson, J. (2017, March 16). What is fake news? Its origins and how it grew in 2016. *Telegraph*. Available at: http://bit.ly/2nYV2bI. Accessed April 2017.

Edwards, J. (2016, June). Article Jeremy Corbyn's Facebook strategy is so much more sophisticated than you think. *Business Insider UK*.

Available: https://uk.news.yahoo.com/jeremy-corbyns-facebook-strate-gy-much-072528169.html. Accessed February 2018.

Herman, E. S., & Chomsky, N. (1998; originally 1988). *Manufacturing Consent: The Political Economy of the Mass Media*. New York: Vintage.

Jensen, M. J., Danziger, J. N., & Alladi, V. (2007). Civil society and cyber society: The role of the internet in community associations and democratic politics. *The Information Society, 23*(1), 39–50.

Miller, C. (2016). *The Rise of Digital Politics*. London: Demos.

Parmelee, J. H., & Bichard, S. L. (2013). *Politics and the Twitter Revolution: How Tweets Influence the Relationship Between Political Leaders and the Public*. Lanham, MD: Lexington Books.

Pickard, J. (2016, November 6). When politics and social media collide. *Financial Times*. Available at: http://on.ft.com/2p2ZwOB. Accessed April 2017.

Runciman, D. (2016, October 31). Politics has gone wrong. Is digital technology to blame? *Guardian*. Available at: http://bit.ly/2nM4YFd. Accessed April 2017.

Siddique, H. (2009, July 29). David Cameron says sorry for 'twat' comment during radio interview. *The Guardian*. Available at: https://www.theguardian.com/politics/2009/jul/29/david-cameron-apology-radio-twitter. Accessed February 2018.

Troy, G. (1998). JFK: Celebrity-in-Chief or Commander-in-Chief? *Reviews in American History, 26*(3), 630–636.

6

Children

Eroding Their Childhood?

> Children are exposed to far more from an earlier age because of the internet and this is eroding their childhood. We don't let children be children anymore. The innocence of youth is lost too early in their lives nowadays than at any other time in the past. (Male, early forties, Northampton)

> Children might be more technically savvy that at any point before but they are growing up too quickly as technology permeates every part of their childhood. (Male, early fifties, Derby)

These were two of many responses from our *Screen Society* participants that consistently referred to a new culture of contemporary childhood that primarily centred on the regular and widespread consumption of new media technology via screens. For many children the internet is a world of wonder with endless possibilities to connect, create, communicate, be entertained and acquire knowledge at the end of their fingertips. As soon as they learn to swipe and tap at whatever age they gain access to electronic devices, a child is involved in a process of creative and interactive learning that has limited online boundaries. (Whilst

© The Author(s) 2018
E. Cashmore et al., *Screen Society*, https://doi.org/10.1007/978-3-319-68164-1_6

recognising the onset of puberty through post-puberty period of adolescence, for the purposes of this chapter we follow the definition of a child proposed by the United Nations Convention on the Rights of the Child as "a human being below the age of 18 years".)

Despite this cultural shift, the internet also comes with flashing hazard warning lights of potential dangers that children do not often realise exist until it is often too late. In a 2017 article in the *Telegraph* newspaper, for example, it was reported how a 5-year-old English boy had become the youngest person to be investigated by police for 'sexting' (the process of sending sexually explicit pictures or messages through a mobile device). Here it was revealed how the number of cases of children taking explicit pictures of themselves, commonly those aged either 13 or 14, had soared with more than 4000 investigated by police since 2013 (in the UK it is illegal to take, possess or distribute an image of someone under the age of 18 even if it is the person who took the picture of themselves in the first place). In the same article it was also reported that in a 2013 survey conducted by the NSPCC and Childline, 60% of teenagers had been asked for a naked picture of themselves by another person and 40% had actually taken a naked 'selfie'.

So far this book has highlighted how screens have completely changed our everyday practice, impacting upon our method of communication, relationships, habits and life style. But what about the impact of screens and the internet on children? In his 1991 book *Modernity and Self-identity*, Anthony Giddens illustrates that in a post-traditional order, self-identity becomes a reflexive project that we continuously work and reflect on. Since the mid-1990s children have been born into a digital world and unlike previous generations where this was not available it has become an obvious and important part of everyday life for millions of children across the world who seek to explore and self-express with like-minded others. Not surprisingly, therefore, questions are being asked about how integral the internet is in the lives of children and the impact it has on their social and cognitive development. In addressing this, the focus of this chapter is on the data captured from our *Screen Society* participants regarding the impact of the internet in this new contemporary culture of childhood.

Evolution of Childhood

Childhood has never stayed the same. In the industrial period of the eighteenth and nineteenth century there was an expectation on children to go to work in order to add another financial contribution to the family economy. Legislative changes throughout the nineteenth century sought to change this (including the Factory Act in 1833 which limited the length of a child's working day), but by the turn of the twentieth century the permissible age for child labour across the western world was still as young as 12. As the twentieth century progressed, the increasing focus on children's education, rights and welfare changed the purpose of childhood from economic to emotional. This was particularly pertinent post World War II when the standard of living rose across a number of Western nations, including better levels of family income, housing, health and education standards. It also saw the advent of television, although this initially led to concerns surrounding its passive consumption and the perceived cognitive impact it would have on the attention span of children. Towards the end of the twentieth century there was an increasing number of authors (such as Roald Dahl, Enid Blyton and J. K. Rowling) engaging children in fictitious stories, but it was the emergence of the internet and new media from the 1990s that really started to transform the culture of childhood.

In her 2009 book *Children and the Internet*, Sonia Livingstone states how online social networking has completely reshaped the communication practices of children and consequently their self-identity and relationship management with people. Given its constant availability, this can consist of downloading music and videos, creating their own content via blog and channels such as YouTube, or communicating in real-time with one or more people through sites like Facebook, WhatsApp, Instagram, Bebo, MySpace and twitter. In their 2013 article focusing on young people and online social networks, Fatimah Awan and David Gauntlett suggest that the online everyday practice of young people relates to three primary areas: connecting and convenience; openness and control; and privacy and authenticity. Children use social networking sites and instant messaging to maintain contact and communicate with friends and family as well as to create online identities that co-exist with their offline identity. This allows them to exchange information, gain knowledge and new contacts easier and becomes an extension of their everyday social worlds.

In his 2016 article for the *I* newspaper, Adam Sherwin highlights that there had been a 25% decline in children's television viewing, with a structural shift emerging towards tablets and mobile devices to provide the necessary entertainment sought by children. Focusing on YouTube, he stated that out of the top 100 channels based on video views, 35 were aimed at children and generated 8.6 billion views. Likewise, in his 2016 article for the *Guardian* newspaper, Jasper Jackson outlines how 5–15 year olds were spending three hours a day using the internet compared to just over two hours watching television.

Reference to the shift from television to online entertainment for children was a consistent topic of discussion across the data, including this response by a female in her late teens from Glasgow:

> There is relatively little difference between watching television for hours or spending time online. If anything being online (depending on what they are doing) is better than sitting in silence passively watching television. If the child is watching YouTube videos it allows them to read or engage in discussion in the comments. If they are playing online games they interact with the game more than they ever would by watching television.

This increasing focus on active rather than passive consumption was raised by Sonia Livingstone, Leslie Haddon, Anke Görzig and Kjartan Ólafsson in their 2014 book that reflected on the findings of an EU Kids Online survey aimed at children aged 9–16 across Europe. Of relevance to this chapter were the main online activities which included watching videos (86%), communication (75%), downloading and sharing videos (55%) and chatting and/or blogging (23%).

Writing about the 2015 Pew online survey of American teen usage of social media and technology in her 2015 article *Teens, social media and technology overview*, Amanda Lenhart outlines how 92% of young people aged 13–17 are online daily, whilst 87% own a computer, 58% a tablet and 73% a smartphone. The most popular form of communication was through Facebook (71%) followed by Instagram (52%), Snapchat (41%) and twitter (33%). The Pew online survey also found how nearly three-quarters of teenagers have more than one social network profile, thus indicating widespread everyday engagement across a range of social networking sites.

Although internet access is not uniform across all children and young people given that access is often based on a number of factors (such as cultural, social and financial), for those that are part of this new culture, it is particularly the case in the home for those families that have devices that children can access. For children, the internet is a space where they can enhance their sense of individualism from the comfort of their own home. They don't often fall back on the increasing rights they have been given since the twentieth century; instead they feel as if it is their right to be given the freedom to use the internet for their own idiosyncratic purposes. For example, in the research reported by Jasper Jackson above, three-quarters of children that have internet access consume it in the privacy of their bedroom.

In research conducted in 2008 by the European Commission, it was stated how 47% of 15–17 year olds access the internet from the privacy of their home (compared to 22% of 6–10 year olds), 57% of 15–17 year olds access it at school (compared to 49% of 6–10 year olds) and 32% of 15–17 year olds access it from a friend's house (compared to 16% of 6–10 year olds). Whilst the comparison of these findings might be what people expect in terms of higher levels of internet consumption amongst older children, the results still emphasise the informal importance of the home and the formal importance of the school as the two primary sources of internet access.

In fact, the importance of the internet in aiding the learning process of children was raised by a number of participants, including this male in his late forties from Kettering:

> The internet is a brilliant educational tool that is way better than what the education system of a school can offer. Parents can install apps on mobile devices for every subject area being 'taught' and all world knowledge is available online. No longer do we need leftie loon teachers filling kids' heads with nonsense or brainwashing then. I seriously believe we could abolish all schooling.

Whilst this is unlikely to happen, at least in the near future, it does highlight the importance of the internet to a child's education as much as that provided by teachers, books and classrooms. As one male in his early twenties from Ohio stated:

> Engaging with the internet at school and at home is education in the 21st century. The educational establishment have to embrace it or they will be disadvantaging those students that do not access it as the future is clearly going to be more digitally focused.

Changes to children's learning was also raised by Heinz Hengst in his 2001 chapter "Rethinking the liquidation of childhood", where he writes how learning is no longer just the domain for school; instead it is a significant component of everyday life and no longer places children as the ones who need to be qualified and the adults as those who are qualified. Following our own findings on children and their engagement with the internet, we agree with Hengst when he argues that contemporary childhood is going through the process of liberation from modernity's educational project. Indeed, one significant feature of the changing culture of learning for children is the recognition by schools of the importance in online learning. For example, it now comprises a significant component of the curriculum at most schools across the western world, with controls in place to limit what children can access whilst on school premises. However, the private home or at a friend's house is different and relies on the parent or guardians to monitor what sites are being accessed, particularly if online consumption is taking place in the privacy of the bedroom.

One other influential feature that some participants referred to was the role of the media and how it can shape contemporary childhood. Sonia Livingstone raises this in her 2011 chapter "Internet, Children and Youth", when she states that as "young people make the transition from their family of origin toward a wider peer culture, they find that the media offer a key resource for constructing their identity and for mediating social relationships" (p. 348). The comparisons that children make to not only celebrities they see through the media but also their own peer groups was a concern for some Screenagers, including this male in his late thirties from Shrewsbury, who stated:

> One big problem is children comparing themselves negatively to others as this subsequently causes distress. If it is not the image carried by some celebrities it is that their friends received more likes than them on photos or comments on Facebook or through the number of Snapchat or Instagram followers they all have.

Similar concerns were raised by this female in her late teens from Chicago who viewed the internet as a cause of self-esteem problems for children by explaining that boys and girls "almost by default follow "perfect" people on twitter or Facebook and constantly compare themselves to people online."

The Dark Web

In her 2011 chapter highlighted above, Sonia Livingstone refers to how "the internet is structured around a strong tension between two competing conceptions of childhood" (p. 350). On the one hand, children are often seen as vulnerable and the internet poses a risk to their social and cognitive development as well as the potential risk of causing them physical and emotional harm. As a consequence, regulatory protectionism is deemed as a necessity to protect them from the dangers that the internet can bring. On the other, children are seen as naturally creative and the skills that can be developed and enhanced through the internet are underestimated by the adults around them.

For all the good that the internet provides to children it is also important to recognize that it also has a dark side. Although some children have impressive digital skills these are not universal and there are examples where some fail to recognize the risk and dangers posed in what is often a faceless process of communication. Opportunities and risks are inextricably linked and for children to make a new faceless friend or follower online there is the risk that this could be someone who is ill-intentioned. For example, the 2012 chapter examining cyberbullying among college students by Robin Kowalski, Gary Giumetti, Amber Schroeder and Heather Reese reports how one third of young people are contacted by someone they don't know.

Moreover, in the 2014 book referred to earlier by Sonia Livingstone and colleagues, it was found that less than 10% of children aged 9–16 had actually met online contacts in an offline environment. This was also a consistent finding within the data, echoed by this female in her early twenties from Leicester: "Children are now at far higher risk of gaining unwanted attention through anonymous online forums and other social network sites", whilst this male participant in his late teens from Madrid stated: "If you look at my twitter or Instagram followers I reckon I actually only 'know' about a tenth of them, with the figure even lower in terms of ever meeting them face-to-face in a social situation."

Ulrich Beck's 1992 book *Risk Society* that outlined a shift from a modern industrial society to a late modern society has particular theoretical pertinence to the relationship between the internet and childhood. As this chapter has outlined so far, the internet has provided increasing opportunities for children to present themselves as individuals within their own specific patterns of online consumption. Beck explains that situations like this can create a risk society that becomes "a systematic way of dealing with hazards and insecurities induced and introduced by modernisation itself" (p. 21). For Donna Chu, in her 2016 article that focused on internet risks in Hong Kong, risk is a contested concept because it is not the same as harm, but instead refers to the probability of certain events happening in the future.

One element of the presentation of a risk society is through the representation of news by the media. When this concerns children it often leads to a heightened form of everyday public and parental anxiety regarding child safety. This has resulted in a social consciousness often influenced by stories in the media concerning elements of the internet that we do not want children exposed to. For some Screenagers, like this male in his early thirties from Edinburgh, exposure to the media is an important feature of childhood because it provides an opportunity to reflect and engage with the way that news is represented:

More interactivity with their chosen media source helps generate a sense of thick skin towards what could be perceived as negative media influences and this provides a range of educational benefits and greater opportunities to become involved in debates, both online or in person with stories being presented by various media outlets.

Despite the presence of views like this within the data, for a number of parents, teachers, the media and child protection experts, there is a constant focus on child well-being and safety concerns surrounding ineffective privacy protective controls on some sites, particularly when it subsequently leads to bullying, harassment and grooming as well as other forms of inappropriate content and communication. This has the potential to cause significant emotional and physical harm to people and has led some site providers to develop technical strategies to minimise risk including attempts to ban children under a specified age, to restrict social networking to children only or to install buttons to report any abuse or concerns about their safety. But a key question remains: how efficiently are they policed? With limited regulation on the internet, age can be disguised to make a person act younger than they really are.

The accessibility of children on the internet has led to online grooming as it provides an environment for sexual predators to anonymously operate within the confines of their privacy that was not available twenty years ago. Grooming usually involves the manipulation of children by initially striking up some form of communication before encouraging them to engage in some form of risky behaviour such as arranging to meet face-to-face at a specified location that then has the potential to lead on to sexual and/or physical abuse (in the UK the legal age for consensual sexual activity is 16). However, the 2010 article examining the digital literacy of Flemish youth by Sofie Vandoninck, Leen D'Haenens and Verónica Donoso found that most young people are sufficiently skilled on the internet that they devise strategies that protect them from potential harm.

Examining another dark side of the internet, in her 2007 book *Cyberbullying and Cyberthreats*, Nancy Willard developed a taxonomy to explain the many different forms and venues in which cyberbullying takes place. In this taxonomy she included harassment (repeatedly sending offensive messages to a person), exclusion (blocking a person from shared groups), cyber-stalking (stalking another person by sending repeated threatening communications) and impersonation (posing as the victim and sending negative or inappropriate forms of communication to other people or groups).

Although bullying predates our attention to screens, cyberbullying has become a concerning feature of everyday practice for some and has moved the power from face-to-face confrontation to what is now also communicated through an electronic device (such as via email, chat rooms, text messages on a phone or instant messages on social media sites like twitter, Facebook and Instagram). In their 2013 article, Sonia Livingstone and Ellen Helsper outline how the internet and other digital technologies have changed the opportunity and risk taken by young people. Whilst the opportunities were raised earlier in the chapter, for some participants the risk is not always purely online, but also through the constant electronic surveillance in their offline everyday life. As explained by this male in his early forties from Portsmouth: "Children cannot make mistakes in the playground because someone will tape it and upload it on YouTube for millions to watch. It's easier to manipulate and take advantage of the weak, vulnerable or less educated."

Of course, all risks can be balanced against the opportunities that the internet brings for children. Whilst recognising the dangers posed to those children who are online, this male in his early twenties from London stated:

> In the same way people were finding issue with the invention of television, the internet is obviously being considered as an unfiltered stream of raw content. There are areas of the internet that are obviously completely unsuitable for children but for every awful site, there are several that educate children more than television can do.

However, other Screenagers were aware of the risks that the internet poses to children and many thought they were more pronounced than they ever were with television. As one male in his late twenties from

Glasgow stated: "Even with parental controls, Google can lead to too much truth. Some things children should be shielded from a lot better. As much as parental controls can stop children accessing dirty websites, it won't stop them googling things", whilst this female in her late forties from Indiana shared her worrying signs emerging from her son's use of the internet:

> My son is watching things online that I don't think are age appropriate for a 10-year-old, and he uses some slang that tends to be overtly sexual. He has no idea what sex is. Up until a couple of weeks ago, he didn't realize girls do not have a 'weiner'.

With the increasing focus on sex education, widening contraceptive use and sexual experimentation at a younger age it is fair to assume that the internet is only likely to stimulate this even more in the future, particularly given the unfiltered availability of pornography to those who type it into a search engine.

Other Screenagers suggested that children often understand the risks posed by the internet, but part of its thrill is the open navigation it provides. For example, this man in his late teens from Newcastle upon Tyne explains that children will always choose to "interact in ways that are exciting, increasing the risk of exposure to extreme imagery that is pornographic or violent in nature", whilst this female in her early thirties from Toronto shared similar thoughts:

> The internet is largely unregulated and a lot of its content makes a mockery of our laws. For example, minors can easily and quickly access pornography – whereas traditionally that would be very difficult for minors in "real life". There are also a lot of "video nasties" online too, such as videos released by terrorist groups. Again, minors can encounter these very easily. I think the main consequence of all this is both a loss of innocence and the significant degrading of a person's ability to empathise with others. If a youth learns about sex chiefly from pornography, or is exposed to a lot of violent videos, then I do not think they would have very healthy attitudes, or be a very sensitive person in later life.

Parental Controls

What often starts with young children learning to be entertained online (either through watching or by playing games) can quickly become more sustained explorations across the internet. In seeking to develop an identity with peers or significant others, some children quickly learn how to share an image of themselves, but don't often understand the significance and the consequences this action can have on their well-being (particularly if this image is of them naked or exposed in some way). The problem for parents was raised by this male in his early thirties from Doncaster:

> The internet is far more difficult to regulate than television. Parents face a very difficult and time consuming task in monitoring what their children do online. The probable consequences of this are an earlier sexualisation of children than at any time in the past.

Given examples like this and those raised earlier with regards to sexting, if children have online access to unfiltered sexual material, they seem to want to use it and in some cases share it—even if they don't understand it. Although part of childhood is learning through errors that are often made in the absence of emotional or intellectual maturity, in the process of sexualising themselves, once they hit 'send' the distribution and consequences are irreversible and potentially very damaging.

In his 2010 article looking at the regulation of young people's access to the internet, Thomas Wold discusses the assumption by adults that children lack maturity and the internet is therefore a threat to childhood innocence. Given the pleasurable nature of the internet for children a balance is required between the right to privacy and the need for safety. Indeed, some participants sought to emphasise the role of parents in seeking to control the usage of some sites by children. "Parental controls need to be tighter … every parent likes to trust their children but maybe we are more trusting than we should be. Bullies and grooming are the most likely consequences of being too trusting", said this male in his late forties from London.

Part of childhood is about rebelling against parents and, in some cases, society, and children often carry on regardless even though they might be aware of the potential risks to their safety. They also crave independence from parental supervision, as raised by Karen Bradley in her 2005 article examining adolescent development on the internet, where she stated how the internet allowed children to experiment in a space that was unmediated by adults. Whilst children are at the forefront of this social and cultural change, adults worry about the vulnerability and exposure of children. Some of the advice given to parents is to update the software on their devices within online access, change passwords or familiarise themselves with the security settings of various online sites that their children are using.

Reporting on the availability of spyware for parents to use in placing their children's online habits under hidden surveillance, Katie Roiphe in her 2013 article "Would you spy on your teenager?" states:

> The argument is that soon we will be living in a universe where everyone puts spyware on their children's computers, and those of us who resist are fruitlessly clinging to the old world, not to mention sentimental notions about the boundaries between people and the tantalising promise of privacy.

Not surprisingly, the issue of parental supervision was a discussion point for many Screenagers. "It depends what they are doing online. Any good parent will monitor their time spent online and check what they are watching on YouTube, who they are communicating with and what sites they are engaging with", said this male in his late forties from London, whilst this male in his late thirties from Verona stated how the technological expertise of some young people impacts on the potential for parental supervision: "The main concern for me would be the inability to successfully screen what information my child is accessing online if they can cover up what they are accessing. This worries me no end."

Through examples like this, it is clear that the pace of the internet's diffusion into various virtual spaces has, for some adults, outpaced their ability to adjust and manage their child's use of the technology.

In her 2011 chapter referred to earlier, Sonia Livingstone explains this as a "reverse generation gap" where a parent's authority is weakened by the ability of the child to use the technology at a more advanced level than their parents and subsequently avoid adult supervision of the online sites they are consuming. Likewise, in his 2016 article illustrating parenting styles and the impact of this on the internet usage of children, Hasan Ozgür reports how parents often feel powerless if they are not technologically skilled to prevent their children from accessing content that they believe is not appropriate for their age. On this point, the 2011 article by Ana Nunes de Almeida, Nuno de Almeida Alves, Ana Delicado and Tiago Carvalho illustrates how the pattern of parental monitoring of a child's internet consumption is often based on the educational background of the family, with those parents who are educated and digitally skilled more likely to have some form of an interventionist approach with their children's online consumption.

Indeed, some parents establish rules to try and regulate the internet use of their children and attempt to avoid what they see as inappropriate online consumption, including this response by a female in her early fifties from Ohio:

> I don't like doing it, but with teenage twin girls I feel like it is my parental duty to watch what they do online. They might not like it but they do realise I am doing this for their own good (and mine if I am honest). They can go online at times in which we agree as a family, but they have to do it in a room which I am also in at the same time.

Other participants such as this male in his late forties from Sydney recognised the ease with which some parents seek to keep their children entertained:

> Kids should be enjoying themselves doing a variety of things especially outdoor if possible, it sets up habits for life. Unfortunately whereas in the past if parents wanted an easy time they would stick them in front of the television, now they just hand over their smartphone and life becomes peaceful again. You have only got to be in any public space (bus, train, shop, restaurant etc.) to see this sort of behaviour happening.

The Internet Playground

Frequent online engagement by children can lead to suggestions that they are 'addicted' to the internet. Indeed, in their 2013 article "Internet addiction in adolescents: Prevalence and risk factors", Daria Kuss and colleagues refer to how the increasing consumption of new media sites by adolescents detrimentally impacts on their identity formation and brain development, leading to poor academic performance, engagement in risky activities, poor diet, poor interpersonal relationships and self-injurious behavior. They also illustrated how frequent online engagement can lead to depression and insomnia, drug use, alcohol use, obsessive-compulsive disorder and social phobia. Likewise, in their 2017 article looking at internet usage and the psychosocial health of children, Betül Işik and Sultan Ayaz Alkaya state how "uncontrolled and excessive use of internet might lead to internet addiction and cause physical, social, and psychological health problems", including decreased social relationships and interpersonal problems (p. 204).

We've dealt with the more general subject of addiction in chapter four and will return to it when we discuss gaming and gambling in Chapter 9. Among our sample were some participants who raised a number of concerns regarding a lack of social development for children when their everyday practice is heavily influenced by the internet, such as this response by a female in her late thirties from Brighton:

> I feel that social interaction and children's health is suffering as a direct result of too much time spent online. Children seem permanently 'glued' to their devices and appear to communicate with each other electronically even when sat next to one another! The art of communication, basic writing skills and human interaction is gradually being lost and I am seeing this first hand as a parent myself.

As another mother in her early forties from Arizona stated:

> I have teens. They have friends over, but they don't seem to interact with each other. Each child is on their own devices and not actually talking

to each other. Based on this observational experiment I question whether my children actually "know" their friends at all.

Although there were accounts like this of concerns regarding the social and cognitive development of children, others disagreed including this male in his early forties from Peterborough:

People used to moan about children watching too much television, now they worry that they're not watching enough! The consequence is we are producing children who can function in the modern digital world. Therefore, children spending more time actively on social media and less passively consuming television is NOT automatically a bad thing.

This female in her early forties from Inverness went even further:

They have so much more knowledge at their fingertips and are far more advanced than any child who has come before them generationally. They can direct their own learning and skip years ahead without being held back by exams and odd targets set by schools and governments. They are also far more skilled at conducting themselves socially and have been far better at learning the rules of the internet playground than most adults I know.

Reference to the internet playground is interesting because it has liberated children through the opportunities to gain new skills and knowledge and be creative in a form of self-expression not available to previous generations of children.

As we argued earlier, the media play a significant role in presenting a case that the social and transferable skills of children are being lost through their engagement with the internet. For some participants it was important for the child and his or her family to provide a balance to the amount of time spent online, including these views by a female in her early twenties from Leicester:

I think screens can have a positive effect on children. It can reinforce things they are learning, develops a range of skills, provides happiness and be a bonding experience with friends and family. It can obviously cause problems if the child neglecting other areas of their lives such as socializing or homework. I think the key is a healthy balance.

Indeed, this female in her early seventies from Darlington concurred:

Television viewing tends to be a family activity, so it is supervised and maintains family bonds. Alienation is probably a danger if a child's online world becomes more important than their family life. There needs to be a balance to what children consume in their lives but this has always been the case way before the arrival of the internet.

We presume this participant's reference to "alienation" means that she felt children experience a state of depersonalisation or loss of identity as a family member.

As this chapter has outlined, the culture of childhood has changed but the media, parents and guardians are increasingly being positively and negatively influenced by their understanding of the construction of contemporary childhood. The media are often blamed for blurring the boundaries between adulthood and childhood and some children get insights into adult topics before they can emotionally cope and deal with them. The result of this are the frequent debates about the innocence of children and the need to protect them when they operate online, but for children the internet provides a form of stimulation to present themselves as autonomous individuals. One consequence is an overly protective approach from parents and some internet sites although some children are so technologically skilled that they can easily avoid any attempt at covert surveillance. The wider ramifications are still emerging given the speed of internet consumption and the accounts from Screenagers leads us to wonder whether we might be witnessing the disappearance of childhood as we had previously known it. Society used to refer to how children should be seen but not heard; now they should see but not be seen and, as a consequence, childhood as it was traditionally understood could be the biggest casualty of society's increasing obsession with screens.

References

Almeida, A., Almeida Alves, N., Delicado, A., & Carvalho, T. (2011). Children and digital diversity: From 'unguided rookies' to 'self-reliant cybernauts'. *Childhood, 19*(2), 219–234.

Awan, F., & Gauntlett, D. (2013). Young people's uses and understandings of online social networks in their everyday lives. *Young, 21*(2), 111–132.

Beck, U. (1992). *Risk Society—Towards a New Modernity*. London: Sage.

Bradley, K. (2005). Internet lives: Social context and moral domain in adolescent development. *New Directions for Youth Development, 108*, 57–76.

Chu, D. (2016). Internet risks and expert views: A case study of the insider perspectives of youth workers in Hong Kong. *Information, Communication & Society, 19*(8), 1077–1094.

European Commission. (2008). *Towards a safer use of the Internet for children in the EU-A parent's perspective*. Brussels: European Commission. Available at: http://ec.europa.eu/commfrontoffice/publicopinion/flash/fl_248_en.pdf.

Giddens, A. (1991). *Modernity and Self-identity*. Stanford, CA: Stanford University Press.

Hengst, H. (2001). Rethinking the liquidation of childhood. In M. Du Bois-Raymond, H. Süncker, & H. Krüger (Eds.), *Childhood in Europe* (pp. 13–41). New York: Peter Lang.

Işik, B. & Alkaya, S. (2017). Internet use and psychosocial health of school aged children. *Psychiatry Research, 255*, 204–208.

Jackson, J. (2016, January 26). Children spending more time online than watching TV for the first time. *Guardian*. Available at: https://www.theguardian.com/media/2016/jan/26/children-time-online-watching-tv.

Kowalski, R., Giumetti, G., Schroeder, A., & Reese, H. (2012). Cyberbullying among college students: Evidence from multiple domains of college life. In C. Wankel & L. Wankel (Eds.), *Misbehavior Online in Higher Education* (pp. 293–321). Bingley, UK: Emerald.

Kuss, D., van Rooij, A., Shorter, G., Griffiths, M., & van de Mheen, D. (2013). Internet addiction in adolescents: Prevalence and risk factors. *Computers in Human Behavior, 29*, 1987–1996.

Lenhart, A. (2015). *Teens, social media and technology overview 2015*. Pew Research Center. Available at: http://www.pewinternet.org/2015/04/09/teens-social-media-technology-2015/.

Livingstone, S. (2009). *Children and the Internet*. Cambridge: Polity Press.

Livingstone, S. (2011). Internet, children, and youth. In M. Consalvo & C. Ess (Eds.), *The Handbook of Internet Studies* (pp. 348–368). Oxford: Blackwell.

Livingstone, S., & Helsper, E. (2013). Children, internet and risk in comparative perspective. *Journal of Children and Media, 7*(1), 1–8.

Livingstone, S., Haddon, L., Görzig, A., & Ólafsson, K. (2014). *EU Kids Online II: A Large-Scale Quantitative Approach to the Study of European Children's Use of the Internet and Online Risks and Safety.* London: Sage.

Ozgür, H. (2016). The relationship between internet parenting styles and internet usage of children and adolescents. *Computers in Human Behavior, 60,* 411–424.

Roiphe, K. (2013, November 1). Would you spy on your teenager? *The Financial Times.* Available at: https://www.ft.com/content/08887b7a-41c1-11e3-b064-00144feabdc0.

Sherwin, A. (2016, July 11). Children's UK TV programming hits a low as YouTube lures youngsters, Ofcom finds. *I.* Available at: https://inews.co.uk/essentials/culture/television/childrens-uk-tv-programming-hits-low-you-tube-lures-youngsters-ofcom-finds/.

Telegraph. (2017, July 11). Boy, five, is 'youngest person in Britain' investigated by police for sexting. *Telegraph.* Available at: http://www.telegraph.co.uk/news/2017/07/11/boy-five-youngest-person-britain-investigated-police-sexting/.

Vandoninck, S., D'Haenens, L., & Donoso, V. (2010). Digital literacy of Flemish youth: How do they handle online content risks? *Communications, 35,* 397–416.

Willard, N. (2007). *Cyberbullying and Cyberthreats: Responding to the Challenge of Online Social Aggression, Threats, and Distress.* Champaign, IL: Research Press.

Wold, T. (2010). Protection and access: To regulate young people's internet use. *International Journal of Media and Cultural Politics, 6*(1), 63–80.

7

Trolling

Be Kind

> Trolls are a joke in most forms. They only do it [offend people] as banter
> and are just looking to annoy people. So, when the media makes them
> out to be some sort of bogeyman, most of the time that's what they
> want – although I would say stuff like threats of death and rape have no
> place in society. But nine times out of ten, the people that write these
> things, when confronted face to face, would be too scared and weak
> to actually do anything and would, to put it bluntly, shit themselves. I
> don't think troll culture is helped by the fact that, in this day and age,
> everyone is way too easily offended. (Male, early thirties, Aberdeen)

Not everyone would agree with this Screenager. Take Harry Styles, a
known admirer of Gucci: he wore the Italian designer label's suits, loaf-
ers and shirts. It was one such shirt that landed Tess Ward in trouble
in Spring, 2017. *The Naked Diet* author Ward posted on Instagram a
picture of herself wearing a baroque blazing red floral shirt identical to
one owned by Styles. It was all the confirmation Directioners needed
that she and Styles were an item (two of us believe there is no need to
explain that Directioners are devotees of the boy band One Direction,

© The Author(s) 2018
E. Cashmore et al., *Screen Society*, https://doi.org/10.1007/978-3-319-68164-1_7

but the third insists we should elucidate. While we're elucidating, the shirt cost £725).

Ward must have wondered how her life would have been different if she'd have resisted the temptation to post the picture and, instead, hit "delete." "She became a 21st-century Yoko Ono, loathed by loyal Directioners, who are notorious for making voodoo dolls and sending death threats to any woman with whom their idols socialise," wrote Julia Llewellyn Smith, of *The Australian.* (Yoko Ono was John Lennon's partner from the mid-1960s to his death in 1980; many Beatles fans thought she was responsible for the breakup of the band, smearing Paul McCartney and having a negative effect on Lennon's music—there was no internet then, of course.)

The backlash against Ward started with a sudden proliferation of 1★ ratings for her cookbook on Amazon. That was just preamble: the merciless trolling on Instagram forced Ward to beg her new followers to "be kind." Of course, this gave the Directioners impetus. They were—and probably still are—especially remorseless fans. No fan welcomes love interest such as this. Social media platforms like Instagram provide more access to celebrities than at any time in history: fans can track their idols' day-to-day movements and perhaps even get a "like" or a reply. There's a quid pro quo: fans feel a sense of entitlement. So when a pop star like Styles starts a relationship with someone, it will become known immediately. That much is taken for granted. But, when the new partner flaunts online, she might as well be rubbing the fans' noses in it. Earlier, in 2016, when it became known that One Direction's Louis Tomlinson's partner, Briana Jungwirth, was pregnant with his child, she was bombarded with an unstoppable online invective that included death threats.

Ordinary People

Unlike in real space where no one can hear you, in cyberspace, everyone can hear you—at least potentially. They can also read and see you. In a sense, that's the beauty of the internet: no one edits and everyone is free to say and write whatever is on their minds, offend or please. The net has commissioned many, many freedoms to explore, learn, exchange and form allegiances. But it's also enabled a more sinister capability.

While the Style-Ward Instagram storm raged, a less conspicuous, though more astonishing incident concerned a 23-year-old Royal Air Force gunner who disappeared after a night out. The working theory was that he had fallen asleep in a rubbish skip that was later taken and emptied in a landfill site, while he remained unconscious. Understandably distressed, his mother Nicola Urquhart talked to the media about her torment. This was the greenlight for media users who took to the internet to post allegations about a family cover-up operation and miscellaneous theories purporting to explain the vanishing. The sinister capability, as we called it, manifested in accusations that left the family in distress. "I'm at a loss to see why they do this," Urquhart admitted to Thomas Burrows, of the *MailOnline* in 2016. "I'm stunned."

Some would say she was naïve and shouldn't have been surprised, let alone stunned. Screenagers are at liberty to post whatever they want—not just within reason, either. Yet they continually post messages designed to limit the liberty of others. This is not so much an irony as an inversion, a reversal of the intended order of things in which ostensibly free people are at the mercy of implacable forces. For a supposedly open environment, the internet can be oppressive. What *is* genuinely ironic though is this: people who might, in normal circumstances, be expected to be fair, reasonable and friendly, once in an online community, deliberately behave in ways that are totally contrary to expectations.

Trolls were, in folklore, unpleasant though not especially malicious cave-dwellers; today, they are deliberately offensive characters, who use the net to upset and elicit angry responses from just about anybody they choose. Upset might be understating their motives too: those who issue death threats clearly intend to terrify their targets. What makes this even more frightening is that some researchers reckon anybody— literally anybody—is potentially capable of this kind of behaviour.

Conventionally, trolls are considered marginal though utterly dislikeable and unwanted presences online. "Some studies even showed that trolls have predisposing personal and biological traits, such as sadism and a propensity to seek excessive stimulation," explained Cristian Danescu-Niculescu-Mizil, Justin Chengg and Michael Bernsteinn, of Cornell University, who in 2017 recruited 667 participants through an online crowdsourcing platform to try to understand the origins of online abuse. Perhaps at this point, we should illustrate the type of studies they have in mind.

According to a 2014 study entitled "Trolls Just Want To Have Fun" by Canadian academics Erin E. Buckels, Paul. D. Trapnell and Delroy L. Paulhus, trolls operate as "agents of chaos" exploiting hot-button issues for perverse pleasure. "Online trolling is the practice of behaving in a deceptive, destructive, or disruptive manner in a social setting on the Internet with no apparent instrumental purpose," according to the report. For Buckels and her colleagues, trolls are a specific group of people with clear identities and a sense of purpose who exhibit traits known as the "Dark Tetrad" of personality disorders: narcissism, Machiavellianism, psychopathic and sadistic tendencies. They share many characteristics of the classic "joker villain: a modern variant of the Trickster archetype from ancient folklore."

This is a colourful conception of trolls, all of whom are in some way dysfunctional and borderline sadists. Identifying a particular group and cramming them in a separate compartment seems to be one of those easy-on-the-intellect ideas that allow us to think trolls are different from the rest of us. What makes the research of Danescu-Niculescu-Mizil and colleagues disturbing is its opposite implication. Trolls are not a distinct, exclusive group of ill-meaning individuals whose mission in life is to bring misery to others. They are more like gremlins lurking inside all of us, waiting for the right circumstances and then springing into life. "Ordinary people can, under the right circumstances, behave like trolls." (Exactly what the researchers mean by "ordinary people" isn't clear: maybe, as we'll suggest later, trolls actually are "ordinary people".)

Several Screenagers arrived at broadly similar conclusions. For example one said: "A small percentage are really evil their ultimate aim is to cause upset dismay and mental harm. The rest probably don't realise they are 'trolling'." Another believed, "the media call anyone who disagrees or insults someone online a 'troll'." The implication here is that the word is a convenient label to slap on an amorphous group of users who occasionally behave insultingly or even offensively. Both views dovetail with Danescu-Niculescu-Mizil's research. So, in a way, does this, the response of one of our participants, who, when reacting to a question on trolls, replied: "That really depends on the definition of a 'troll'… One who merely posts distasteful material on a timeline might be seen as a fool; however, a former partner making threats to injure another might be seen as a menace."

Amanda Levitt also found the differences between everyday users and trolls were typically exaggerated. In her 2016 thesis for Wayne State University, she concluded, "the key differences found between trolls and other commenters centered around use of language" (p. 59). She investigated abusive messages about fatness. The cyberbullying on fatness (a term she favours, and which we discussed in Chapter 6) reinforces the "dominant discourse" that can be found in any traditional media. In other words, the assumptions, the attention and the way blame is apportioned is the same as in mainstream narratives. The only difference is the language, which is more overtly nasty. But again, the point is underscored: trolls are probably not a distinct group of people dedicated to making life hell for others. More likely, they are Screenagers, who, for potentially any reason, decide to use words and expressions that are deliberately offensive. They may be serial abusers, or they could do it once then never again.

So, if this style of research is to be accepted, the circumstances rather than the individual merit most attention. What are they? On this question the research is not helpful; so we'll try to contrive an answer. We'll start with the internet itself. All of us—including readers and writers of this book—have the capacity and potential to be trolls. It may offend many to learn they are capable of deliberately inciting, offending or abusing others (we'll explain the differences later). And the argument that, in certain circumstances, anybody can tap out sexist and racist words sounds outrageous. Name-calling, profanity, hate-filled messages, sexual menaces, violent threats: they all circulate online, whether on Facebook, Reddit, Snapchat, forums, or anywhere else a user can post a readable message.

The freedom of the internet has obviously been a factor in its extraordinary growth. And it *is* extraordinary: there is no historical precedent with which we can compare it. Not even the steam engine. The lack of regulation has made it attractive, but repulsive. Free speech means exactly that: an unrestricted right to express opinions, views, without censorship or restraint. Americans enshrine this right in the first amendment of its constitution. As such, it's inviolate: it can't be contravened. The principle is less securely protected in Europe and elsewhere, but generally respected.

The objection to free speech is well-documented: it can encroach on freedom. At first, this sounds like a contradiction. But consider: someone posts an honestly felt message expressing his or her extreme dislike of black people. Are black people affected? No. At least, not initially. But, given the unregulated character of the net and the rapidity with which memes are copied and spread with slight variations, the message goes viral and is read by millions, many of whom endorse it. Many others reject it. Many more object to it and challenge the sentiments behind it. (Memes refer to all cultural elements that can be transmitted from one individual or another by imitation or other non-genetic means.)

In this scenario, similar messages proliferate, making a great number of black users feel uncomfortable, if not threatened. The initial message might have expressed dislike, but hate, detestation and outright hostility feature in many posts, creating a rancorous environment. Maybe there are no explicit threats and the hostility could be diffuse rather than directed at anyone or any group, but people feel pressured or intimidated all the same. Is it still free speech?

Many Screenagers believe the free and ungoverned character of the internet is vital and, as such, should be protected, no matter what the consequences. "If it isn't illegal, then censorship is wrong [and] we should be free to express ourselves how we see fit," suggested an Australian woman in her twenties. The responsibility for exercising discretion lies with the user: "That's all on the viewer, on what they're watching." In other words, *caveat emptor*: the consumer alone is responsible for checking the quality and suitability of the material he or she reads. "It is a potentially dangerous environment for viewers," concluded one man in his thirties. Let's discover exactly how dangerous—potentially.

Shouting at Thin Air

How many people are on the receiving end? One in 25 said they are singled out for abuse "all or most of the time," reported Hayden Smith, of the *Independent* in 2016; that's only four percent. Other studies put the number of people subjected to offensive or threatening messages online at any given time as high as 24%, suggesting over-reportage, under-reportage or unreliable research findings. (Catherine Blaya and Michael Fartoukh offer a summary of a number of studies in their 2015 article "Digital Uses, Victimization and Online Aggression".)

The Hayden Smith article (mentioned above) included an interview with Childnet Chief Executive, Will Gardener who revealed that as many as one in four teenagers suffered hate abuse online. "Gender, sexual orientation, race, religion, disability or transgender identity" were identified as the key features for abuse by, though disability and nationality were also named. All studies conclude some kind of abuse goes on continually and that the reasons are arbitrary. Of course, defining abuse is tricky.

Researchers used to use the term "online hate," which could be offensive, mean or threatening, and either targeted directly at a person or group, or generally shared online. In its most extreme form it can break the law (in 2016, figures showed convictions for crimes under a law to prosecute internet trolls had increased eightfold in a decade). Now, trolling is the term of choice. Trolling is often used interchangeably—and often confusingly—with cyberbullying or online bullying; we will stick with trolling, while remaining mindful that trolls does not describe an identifiable group of nasty people who dedicate their online lives to making life unpleasant for others. It may, in fact, *does* appear that way; but we are all capable of causing discomfort to others.

The behaviour is not as new as many assume. In 2006, for example, a group of male teenagers filmed the sexual assault of a sixteen-year-old female with learning disabilities in a suburb of Melbourne, Australia. The video was posted online; it showed the assailants urinating on the girl, setting fire to her hair and forcing her to engage in sex. The victim had met her attackers in an internet chatroom.

This was a physical assault. Other forms of trolling are executed remotely. At the blandest level, trolling can involve spreading lies, usually malicious lies, about someone; the object of the exercise may be to cause the subject embarrassment or annoyance, but it may also be to force someone to suffer more acute forms of discomfort and pain. There have been several cases of suicides resulting from sustained trolling. Forums and online dating sites are the kinds of environments favoured by trollers who enjoy this, though twitter and other microblogging sites are also favoured. The environment is less important than the reaction: the purpose of an inflammatory untruth is to provoke. If the troll can prod someone into responding angrily, that's fine. If there is silence, then the troll also takes satisfaction from the knowledge that the object of his or her remarks is either intimidated or too upset to answer.

But Janice Turner, of the *Times*, disagrees. Writing in 2015 she states: "When I hear that girls feel silenced on social media because their every utterance is greeted with variations on 'shut up, bitch', I think, job done for the trolls!" Turner is writing specifically about what she calls "online misogyny," which is, as the term suggests, abuse directly exclusively at women, though she also understands that trolling "encompasses threats of rape and violence, or persistent harassment that borders on stalking." And these, of course, can be directed to both sexes. By both.

Likewise in 2014 Emma Barnett, of the *Telegraph*, acknowledges: "The majority of trolls are men. No doubt about it." But she goes on, "there are women behaving like this too." Sometimes women attack each other online to defend men. "The internet has effectively made them [trolls] disconnect their sense of morality from their actions."

To complicate matters somewhat, there is also the practice of self-trolling which, as the name implies, involves posting toxic messages about oneself. The primary purpose of this is to pre-empt other trolls. A secondary purpose is to express self-loathing. While there are studies that confirm that as many as one in ten internet users have admitted to posting nasty messages about themselves anonymously. This could be interpreted as a new way of self-harming, which itself is a response to many different forms of anxiety and moods, often initiated by a frustration with one's own physical appearance.

So, what do we know and what don't we know? We know for sure that forms of hate or abusive comment or deeds have been circulating on the net practically since the advent of the second stage of development of the internet, when the net itself became universally popular. It wasn't always known as trolling and it wasn't conducted exclusively online, but the motivation was much the same: to make others suffer. We also know that the net itself is an environment that commissions this type of abuse: not by design, of course; but the very nature of a relatively ungoverned and, in practical terms, ungovernable space in which communications are allowed to flow freely is ideal for those who wish to perpetrate hatred and justify it in the name of freedom of speech. And, while we suspect online abusive ebbs and flows somewhat, "most research shows that the rate of cyber-bullying has held quite steady across all platforms," for several years, according to Aman Jain (reporting in 2016).

We know there's a lot of bile spilt on the net, but we can't quantify with any accuracy how much abuse there is. Scholarly studies produce hugely different estimates based on the recipients' accounts and, because there is no universally agreed-on definition of hate, abuse or offence, these accounts are necessarily subjective. There's nothing wrong with subjective experiences as research data, of course. But it does tend to vitiate attempts to produce a description or reliable report that's valid across time and space. Obviously, the number of people who consider themselves trolls and consciously try to impose their unappetizing messages on others is also unknown—though probably not unknowable. Our reason for supposing this is that those who identify themselves as trolls are not like, for example, a burglar, a violent offender or an anonymous writer of malicious mail: all these characters prefer to stay outside the purview of the authorities or general public. Trolls tend to be quite proud of their offences and welcome the attention afforded them, especially by the media. This observation is based solely on one instance, but it's worth recording.

Trolling as Power

In 2012, one of the present authors was invited onto a television news programme to discuss the case of a teenager with Down's Syndrome, who had been trolled many times. Photographs of her had been taken from a website and circulated. Asked what kind of person would wish to perpetrate such abuse, the author could only surmise: "The kind of person who is not typically given much respect or authority in other aspects of his or her life. They would probably not have a job with much seniority or power and, as a way, of compensating, assume power online." The interviewer seemed surprised: "Power?" "Like it or not, there is a certain empowerment in trolling: it means you have the power to create unhappiness and perhaps acute distress in the life of another human being. It's a perverse power. But it's power all the same," was the response.

The trolls had a new target: within hours, the author was in their sights. Among the mostly angry and sarcastic put-downs (some quite humorous) were serious comments about the legitimacy of trolls. Several trolls reminded the author of the apparent sanctity of free speech: "Free speech is just that: free. It means that just because most people are free to say nice things don't mean everybody has to. I choose to use my freedom of speech differently: to be nasty to other people." This paraphrases one of the trolls' justifications: it isn't word-for-word accurate, but it captures logic in the seeming loathing. (The story of the teenager with Down's syndrome is at: http://bbc.in/2rmcrrT.)

It also chimes with a comment from one Screenager: "Your annoyance may be someone's truth." The sense of this isn't immediately apparent, but, presumably annoyance is taken to be the state of being annoyed or irritated; the participant regards this is only one among several feelings evoked by a message. The sender (in this instance, we'll assume it's a troll) is making a remark that he or she takes to be true. It is hurtful and provokes vexation in another or several other parties. But it is the sender's right under the rubric of free speech to make the remark, regardless of its consequences. Again, there is logic, even if it is not a logic most people would wish to recognize. "I personally believe that if you put yourself up there in the social media world sadly you have to accept some of this kind of crap," concluded one participant.

There are few consistencies among Screenagers about trolls and trolling. But one is that they are mostly harmless and worthy of ridicule rather than the fear they presumably wish to incubate among users. But there are exceptions. Consider this response by a mature English male in his late fifties:

> A large part of the community are more mentally robust and can accept trolling for what it is - a mindless person sitting at a keyboard spouting bile at anyone who will listen. A bit like the street 'nutter' shouting at thin air in the street. If no one listens, it has no meaning. However, a message sent to an individual could be construed as a private conversation between two people, which in most trolling cases it isn't.

Another participant, this time female reacted to the word troll:

> If you mean someone who posts inflammatory content hoping to get people to rise to their bait, then they're harmless and probably do some good in creating interest in a topic. If instead you mean people who post hate comments, then they're obviously 'a bad thing'.

But, it seems, not dangerous to most: "Most people just ignore them as a minor irritant - and sometimes they can be very funny." Most, but not all; as this remark reminds us: "I think they may be a threat if it's a prolonged attack on a vulnerable person (e.g. a child or teenager), but for most they are just an annoyance," said a man in his twenties, echoing the sentiments of a great many participants. Another man, in his thirties, summed up: "They [trolls] can be a laughing stock, but in certain circumstances they are menace, for example with vulnerable people."

The import of these representative views is clear: for the most part trolling can be readily assimilated by Screenagers as deliberately offensive and designed to occasion revulsion, but worthy of only "annoyance" or laughter ("they can be very funny"). Yet they can be genuinely hurtful for a small minority of Screenagers, who are, for some reason, vulnerable. Their vulnerability may be a product of their age, though the participant quoted earlier who considered teenagers vulnerable has seriously underestimated the shrewdness and practical knowledge of teenagers. Or their vulnerability may be due to a great many other factors impacting on his or her ability to fend off critical and provocative remarks that are intended to wound. (Lesbian, gay, bi and trans—LGBT—individuals tend to be targeted offline, though this isn't reflected online, according to our study.)

Abusers themselves justify their behaviour by explaining victims are not arbitrarily selected: "I used my anonymity to call out self-aggrandising politicians by reminding them of policies and misbehaviour they would like to forget. There is a huge difference between that and bullying a teenager on Facebook."

So how does someone who victimizes others on Facebook justify the action? Celebrities are fair game to practically everyone. Even a death threat against a celebrity isn't necessarily regarded as serious: it goes with the territory. Celebrities are, by definition, attention-seekers and deserve bad as well as good attention. Most Screenagers are unapologetic about this kind of abuse. Celebrities almost invite it. Politicians too fall into this category. They typically thrive on publicity. There is also repetition: "If the troll is attacking someone about something they have done or making a comment about an appearance then most of the time it will be harmless and only done once." If it's repeated, then it becomes troublesome.

Nothing and nowhere is safe, of course. That was true before the rise of the internet. As one participant pointed out:

'Abuse' existed before smartphones ... I've heard a lot of people saying that it [abuse] is cowardly as it is behind a screen, implying they would beat the shit out of someone abusing them if it was in 'real life'. But you can reply with words as civilized people would do anyway.

The participant, a man in his thirties from Lisbon, makes several points. His first is that abuse was not created by the net: it is a side effect and has been provided with a new means of transmission. But only the shape or form is new: offensive behaviour, including language predated the internet. His second point is that those who receive abuse online are in a position to defend themselves: all they need to is respond on their keyboards—by hitting the delete button if they wish. They have the means to react, again, if they wish. In other spheres of social life, targets of abuse don't have such means. This hardly amounts to a justification, less still a defence of abusive behaviour, though it reminds us that the near-hysterical response to the supposed rise of trolling may be due to media exaggeration.

In some quarters there is what might pass as sympathy for abusive Screenagers. Even allowing for our previous argument that the number of self-conscious abusers who dedicate themselves to making life hell for others is small ("one percent," according to one participant), that group deserves attention. "The mental health of the trolls," as one participant put it, has rarely been considered. "What drives them to behave in this way? In some cases I'm sure it's immaturity or a lack of empathy but for others it may be indicative of an underlying condition, which needs to be recognised and addressed through the medical system." Another participant concurred, adding:

> One must consider that these methods [online abuse] are being used as a coping mechanism for pre-existing psychological disorders and mental illnesses; so it seems unfair to put the blame for a mental condition on a mechanism that may be helping an individual when really it's an imbalance of chemicals that can't be helped by the individual.

This is a somewhat confused argument, but there is an underlying logic: people are abusive because of a condition that impels them to be nasty and hurtful to others as a way of ameliorating their own ailment. This ailment—if we can call it that—may be the result of a neurochemical imbalance, or similar. This is not necessarily a view we endorse or even find vaguely persuasive; but it is a view of one of our participants, an American woman in her twenties, currently living in Australia.

Sieving the Sea

"Trying to turn the internet into a "safe space" free from all nastiness would be like sieving the sea." This won't stop people trying, of course; but their efforts will still be futile. This resigned approach is typical of Screenagers of all ages. The view finds additional support from a survey of 1500 people aged 14–24, by the Royal Society for Public Health (RSPH), which was published in 2017 as *Status of Mind*. Its conclusion was that, while 91% of the sample used the internet for social networking (an underestimate, in our opinion) there has been a 70% increase in rates of anxiety and depression since 1992. The relationship between the two is postulated—assumed as a basis for reasoning—rather than established. But the research drew links between young people's use of social media, especially when it exceeded two hours per day, and poor mental health. Instagram, which allows its 700 million users (including Beyoncé and Taylor Swift) to share pictures and video clips, was singled out as having the most deleterious consequences. Snapchat, Facebook and twitter were also regarded as harmful.

The researchers believed these platforms can make young users feel worried and inadequate by facilitating hostile comments about their appearance or reminding them that they have not been invited to, for example, a party many of their peers are attending. Would this count as abusive or offensive online behaviour? Probably. Remember, convention dictates that we operate with subjective definitions of harmful practices. The researchers give the example of a woman in her twenties targeted by a group of school peers when she was 15. "Every time I went online I knew I was about to be subjected to a barrage of abuse and hate," she told researchers, adding: "I wanted to know what people were saying about me."

The acknowledgement is probably more revealing than the RSPH realized. The teenage victim felt intimidated to the point where she stopped eating, began self-harming and grew anxious about leaving her own home, but her curiosity overpowered all other considerations and she kept visiting the very platforms on which the abuse manifested. The researchers use the term bullying, which we regard as inaccurate: bullying involves the use of superior strength or influence to domineer and hurt others who are weaker: targets of online abuse do not necessarily lack strength or, for that matter, influence. We recognize, as one of our participants pointed out, online abuse is "a very real problem … to some of the weaker members of our society." Though it is difficult to see how this group could be effectively shielded without subverting the entire purpose of the internet.

The report would find favour with one of our participants who illustrated:

> The internet will always have unsafe areas and cannot be possibly policed without removing free speech. More emphasis is required to teach young people how to use the Internet safely and how to report illegal or nefarious content. It is a potentially dangerous environment for viewers, but equally for an advertiser who could be inadvertently sponsoring an illegal act.

The RSPH recommended that schools teach classes in Personal, Social and Health Education, and for the media to flag up when a picture has been digitally manipulated. As well as Instagram, Snapchat (166 million daily active users) and Facebook (1.94 billion active users every month) are the most popular platforms where users can share photos, videos as well as other information.

"Less is more," cry many Screenagers when the prospect of control, governance and, worse still, "policing" emerge. "I don't see it [the internet] as an 'unsafe environment' nor do I think we are any more 'overprotected' than we are in the 'real' world: it reflects the world," argued one man in his thirties from the north of England. It's an argument consistent with the position of another one of our participants:

> Trolls are simply people being idiots online for a laugh and in reality are quite harmless, they only do what they do due to being anonymous. It's just our current overly touchy PC culture overreacting due to people's inability to take a joke or properly use the block button.

The biggest objection to his would be from victims of doxing. Purists may argue that doxing is fundamentally different from trolling. David M. Douglas, in his 2016 article "Doxing: A conceptual analysis" treats it as a distinct practice. Less puritanical researchers will allow the latter to subsume the former. After all, doxing involves publishing private information about an individual or several individuals on the internet, often with malicious intent. The motivations are varied: a sense of righteousness rather than malevolence could be behind the action; for example, the doxer could be indignant and attempting to expose wrongdoing. As Douglas points out, the doxers, in these instances, want to hold the wrongdoers to account.

It strikes us that doxing is simply a different method of trolling: instead of making snide remarks about physical appearance or other visible features, it involves making information intended to remain private public. The aim is to embarrass certain parties. The term doxing is a distortion of the computer shorthand for documents i.e. docs.

Responses to doxing are much the same as responses to other forms of abuse. On the one hand, users may withdraw, as many of our participants opted to do. For example, one outlined: "I cancelled my Facebook account and will not re-join. It is not pleasant." On the other, intended victims laugh off the abuse and revelations: "If you get worked up/offended by something a stranger says on the internet then you need to have a seriously long look at yourself."

Revenge porn also involves publishing information, though it's difficult to imagine victims not being "worked up/offended" by the prospect of having countless others view what they presumably intended to be private pictures. Revealing or sexually explicit material, when posted online typically for a former partner, without consent, is likely to cause distress or, at very least, embarrassment. That, of course, is the point—and why it's *revenge* porn. The publisher, like doxers, has an axe to grind, but his (and it usually is a male perpetrator) has a private reason rather than a sense of righting-wrongs or just creating havoc.

Our task in this book is not to draw views together neatly in a way that provides an untrue image of life in *Screen Society*: there simply is no single way people conceive of, react to, or philosophize on trolling. Yet there are two questions that crave answers and Danielle Keats Citron asks them on page 55 of her 2014 book *Hate Crimes in Cyberspace*: "Does the Internet bring out the worst in us, and why?"

Horrible People in All Walks of Life

Ginger Gorman, a journalist with the *Sydney Morning Herald*, tells an interesting tale about what began as a routine interview with a super-market employee in 2017 who had all the outward signs of normality. During the interview, he revealed himself as "as a member of a power-ful, international trolling syndicate" who targeted "people with autism and those with mental illnesses." In a frank and repugnant admission, the interviewee mused: "Some people should kill themselves because they're generally pieces of shit."

Gorman recounts how her subject boasted of trolling rape victims, and the Facebook memorial pages of those who died by suicide as well as attacking the page of a young woman who was killed by a train. The story pulses with unremitting hate: the only reward the journalist discerned was the "emotional reaction"—whatever that means (we take it to mean an immediate response characterized by intense feeling to an event or situation, such as seeing offensive or abusive messages or images of oneself). Think about this: the offensive Screenager's only gratification from the extreme abusive he doled out was a certain, barely definable satisfaction derived from arousing intense feelings in others, his victims.

"Trolls are narcissists," concludes Gorman. Again, her terms of reference aren't clear: do they have an excessive interest in or admiration of themselves? It's possible, though the evidence she presents isn't compelling and we found nothing in the *Screen Society* research to support this. Nor did we find any evidence of organized trolling syndicates. This doesn't mean they don't exist, though, with over 2000 contributors to our project, we might reasonably expect there to be recognition of this from someone; there was none.

The twist in Gorman's story is that, following the interview, she became what she describes as a moth stuck in a spider's web—this being the trolling syndicate. Offensive messages started; not necessarily about her, but about minority groups. She found it impossible to escape, at least not without limiting her online activities. Gorman doesn't claim that her contact was or is typical, but the point is: her interviewee seemed reasonably literate, thoughtful in his own way (he said he preferred to target men, not women, rather than be labelled misogynistic), and prepared to explain his motives as best he could. But there is a tendency to monster characters we don't like or, for some reason, find troublesome. By this we mean that we criticize and reprimand rather than try to fathom out why they are so disagreeable. Gorman resisted this tendency and discovered someone she didn't like, but with whom she could communicate in a comprehensible manner; in other words, she made some sense of him and his practice.

We too have to attempt this if we're to address Keats Citron's questions, which we will amend (in italics) in a way that leaves the first question unanswered: "Does the Internet bring out the worst in us, and *if so* why?" "Bring out the worst" usually means allowing someone or perhaps something to express qualities they already possess, but keep hidden or subdued for the most part. So, maybe the internet merely facilitates, but facilitates in an objectionable way. "There are horrible people in all walks of life and all forums," said a young woman in her late twenties from the English Midlands. "The internet just gives these people access to a bigger audience. Some of the stuff is funny, I guess it depends if you're the target or not."

To paraphrase a comment from earlier: "Abuse existed before smartphones." Another participant agreed, his argument being that the sheer volume of people who use the internet has created the impression that there has been an increase in offensive language and targeted mistreatment. In his view, there has been no such thing: the difference today is that we learn about it and are able to react to it quickly.

Responses are as variegated as the forms abuse takes online, but the consensual view of Screenagers is that the net has not brought out the worst in people. It simply provides a different channel of expression, a channel that permits communication with many, many more times the number of people that are typically encountered face-to-face. Poison pen letters, malicious falsehoods, scurrilous rumours and Chinese whispers (in which a message is distorted during its circulation) were all around before the internet, as were more overt forms of physical and verbal abuse, of course. The word bullying in the sense of domineering dates from the late seventeenth century. If anything, as some participants remarked, the internet has allowed a displacement of these into safer domains. This sounds a bizarre conclusion, but victims of more orthodox forms of persecution would probably prefer being subject to trolling, given the choice.

There is also great scepticism among Screenagers about the way in which trolls and trolling are portrayed by the media. We have already cast doubt on the existence of dedicated trolls and favour instead a more fluid conception of trolling as behaviour that's pursued by users intermittently and often with what they feel to be just cause. Examples of this were offered by users who have themselves posted deliberately hurtful and condemnatory messages at specific targets, their intention being to expose lying and what they believe is wrongdoing (particularly among politicians and corporate heads). Several users spoke openly about writing injurious messages about celebrities who have openly courted the limelight. In this case, the justification for oppressing them is basically "live by the sword, die by the sword": if they desire publicity, they should be prepared for all manner of publicity—not just puffery (i.e. exaggerated praise). Few participants in the study objected to this kind of abuse, if indeed we can call it that.

Others consider the choice and status of the target irrelevant or subordinate to the harm either intended or caused. "The consequences of their influence can reach far and the impacts this can have on an individual can be devastating," argued an Australian woman in her forties. "I have heard of personal experiences of people that operate blogs and the effect on their mental state and not to mention their livelihood if they are running a business online can be seriously damaged." Time and again, the word damage cropped up in the responses: "While most [abusive messages] will be dismissed, ignored, passed over or ridiculed, there will be the occasional one that hits its mark and causes serious damage to the victim," remarked a man in his early thirties from London. "I believe trolls have the potential to cause serious damage … the consequences of their influence can reach far and the impacts this can have on an individual can be devastating."

Responses are as motley as the forms abuse takes online: many see people in positions of authority, attention-seekers and other groups as legitimate targets; many consider the detrimental effects of abuse on anyone as too dire. This forces us to wonder whether trolling is, as popular media and nearly half of our 2000 participants believe, dangerous. The alternative is: "They are a laughing stock." That was the succinct view of one participant. Another had a slightly different, if complementary, perspective: "Most of what they [abusers] do is to get a laugh," but added that collectively they can be harmful. "Never underestimate the power of hundreds of internet trolls with time on their hands." A third perspective: "99 percent are laughing stocks, but 1 percent can be serious, depending on the reaction of the person on the receiving end."

This complicates our attempts to answer whether the internet brings out the worst in us, if only because the "worst" may not be as bad as many suppose. If, as a significant number of Screenagers believe, the so-called trolls are not as intentionally harmful nor as deleterious in their effects as many suppose, then maybe they should properly be addressed as figures of fun, laughing stocks, rather like the court jester, the "fool" at medieval court whose job was to poke fun at the noble guests but in a way that stopped short of offending. Trolls clearly do offend some, though perhaps not as many as popularly assumed.

The anonymity afforded by the internet presumably functions as the jester's uniform, which typically included a gaudily coloured suit, hat with bells attached and a mock sceptre. What made his (it was always a man) barbs acceptable was the context: the courtiers and retinue of the sovereign were assembled in the royal household and understood insults would be sharp and so personal that, in other contexts, they would have drawn rebuke. Perhaps the internet community hasn't quite yet come to terms with the grim and cynical mocking of the so-called trolls.

There is no "worst" in us to bring out: the circumstances that form the setting for messages, posts, statements and other kinds of memes are parts of the online context that facilitates, and even sometimes commissions hateful remarks. What's unique about the internet is the sense of knock-on momentum where everything is in motion and every motion affects another motion. A limitless number of possible events leading up to an abusive message will also be factors. As will the consequences: in fact, the consequences are especially important in either motivating the abusive Screenager to troll more, or perhaps lose interest. The meaning of trolling is simply impossible to discern without an appreciation of the context of internet messaging—and this is a context that changes in a blur.

Why? Why not?

In a way, we have also answered the second question, "why?" Most of internet life is ungovernably random yet sometimes appears fatalistically predestined (which is how life generally seems to most of us, anyway). The parts that seem predestined are also the most brutal: weak, vulnerable, helpless, innocent, naïve and people who are easily hurt are most open to attack of some sort. Trolling is the internet at its most primitive: ablaze with taunts and retribution. When we ask why is there trolling, we may just as well initiate the equally challenging inquisition: why is there so much cruelty in the world? And the answer—at least *our* answer—is that, as always, you have to look at not just people, but surroundings, backgrounds, political climates, cultural environments and social conditions. People aren't inherently nice or nasty; situations enliven moods, feelings and behaviour in them.

While this participant wouldn't necessarily agree with our conclusions, he expresses a couple of similar ideas: "If you want to reach out to more people than you normally would, you're bound to encounter more arseholes than you normally would, and they won't be nice to you."

He seems to be saying: life on the net is similar to life, the difference being that the brutality encountered online, though punishing, is answerable with words as opposed to physical violence. People see no real danger on the internet; if they did, they wouldn't go near. The internet is a landscape wild with freedom. It could be tamed. But not without losing the quality that has acted as a lodestone: the net is uncontrollable, anarchic, unruly, disorderly and without law and order.

Arguments collide with force when trolling is the subject. Even among the 2000 participants in our research the field is split in roughly even portions between those who believe trolls are a genuine menace who threaten the peace and mental health of the online world, and those who dismiss them as figures of fun to be ridiculed, not feared. Some believe internet users are under no obligation to be peaceful and are acting quite legitimately in trying to disturb others, either by provoking them into retaliation or by pressuring them into withdrawing from the internet. Most *Screen Society* participants think trolls exist, but many consider them transient; in other words, their involvement in nuisance or abuse is usually short-term. But the practice of trolling itself is permanent: there is always someone who is involved. There is disagreement about the nature of trolling itself: if the target is a character or group in authority or positions of power, then they are legitimate objects of attack, according to some; others insist any type of online abuse is illegitimate.

Our view is that there is a particular kind of interplay catalysed by the internet. The anonymity it provides is both a shield and a licence to wreak damage: users of all kinds of disposition are emboldened to express themselves in ways they might not away from the net. Inhibitions are, of course, germane to social exchange in most areas of life; in fact, much of social life is characterized by constraint. There is also constraint in the cyber-cosmos, though no one loves the internet because of its rules: openness is valued.

Trolls do not exist, at least not in the manner they're popularly depicted. There is not a self-defined coterie that calls themselves trolls and constantly engages in hostile behaviour. There is focused abuse. Users behave abusively for a while and may return intermittently to aggressiveness. So, there is trolling but no trolls.

And the victims? It seems curious that, of 2000 participants in the *Screen Society* project, none described themselves as victims, targets or sufferers of online abuse. Several claimed to know others who had been wounded in some way by unwelcome attention. One participant was an author whose book was, he thought, unfairly criticized online and, although, he counted this as trolling, we tend to think critique or honest disapproval is less severe. There is a serious point here: the claimant may be overly sensitive to censure. Remember, the harm caused by trolling is subjective; this means that the intentions of message senders are less significant. We risk diminishing the hurt occasioned by abuse if we characterize all victims as oversensitive. But there is almost certainly a percentage of those who claim to have been victimized, who, perhaps for reasons beyond their control, are susceptible.

This is not a perspective that will gain unanimous approval among readers. But by 2022, trolling will have been largely consigned to history and people will laugh at what one of our participants called "sad, pathetic individuals who use the anonymity of hiding behind a computer to bully."

References

Barnett, E. (2014, January 7). Twitter trolls: Women are abusers too and we ignore them at our peril. *Telegraph*. Available at: http://bit.ly/2rih1II. Accessed May 2017.

Blaya, C., & Fartoukh, M. (2015). Digital uses, victimization and online aggression: A comparative study between primary school and lower secondary school students in France. *European Journal on Criminal Policy and Research, 22*(2), 285–300.

Buckels, E. E., Trapnell, P. D., & Paulhus, D. L. (2014). Trolls just want to have fun. *Personality and Individual Differences, 67*, 97–102.

Burrows, T. (2016, May 14). Sick trolls target mother of missing RAF gunner Corrie McKeague claiming she knows where he is and is covering up his disappearance. *MailOnline*. Available at: http://dailym.ai/2qyNyfc. Accessed May 2017.

Danescu-Niculescu-Mizil, C., Chengg, J., & Bernsteinn, M. (2017, March 3). Our experiments taught us why people troll. *The Observer*. Available at: http://bit.ly/2pmp30S. Accessed May 2017.

Douglas, D. M. (2016). Doxing: A conceptual analysis. *Ethics and Information Technology, 18*(3), 199–210.

Gorman, G. (2017, June 17). Staring down internet trolls: My disturbing cat and mouse game. *Sydney Morning Herald*. Available at: http://bit.ly/2sK-b4YT. Accessed June 2017.

Jain, A. (2016, October 27). ValueWalk: Twitter trolling: A problem that is just not ending. *Newstex Global Business*.

Keats Citron, D. (2014). *Hate Crimes in Cyberspace*. Cambridge, MA: Harvard University Press.

Levitt, A. (2016). *Crossing the Troll Bridge: The framing of fat bodies on social media*. MA thesis, Wayne State University, Detroit, Michigan.

Llewellyn Smith, J. (2017, May 17). When Harry Styles shared a shirt with Tess Ward. *The Australian*. Available at: http://bit.ly/2qvptG1. Accessed May 2017.

Royal Society for the Protection of Public Health (and the Young Health Movement). (2017). *Status of Mind*, London. Available at: http://bit.ly/2qSwS2t. Accessed May 2017.

Smith, H. (2016, February 9). Safer internet day: 'Quarter of teenagers' subjected to online trolls. *Independent*. Available at: http://ind.pn/2pXXLl8. Accessed May 2017.

Turner, J. (2015, December 19). Women need to man up over online trolling. *Times*, p. 25.

8

Gender

No In-Between

Garfield, the lovable feline cartoon character who rose to fame in 1978 is one of life's laid-back characters. Known to be lazy and gluttonous, Garfield prefers wherever possible, to kick-back and eat pizza and lasagne while watching tv. The characteristics associated with this affable rogue, have led to the widely held assumption that Garfield is male, and yet as one *Screen Society* participant pointed out even Garfield's gender has been debated by internet users. This participant, from the USA in his fifties, cited a URL link to an article from *The Straits Times* that explained more. The article disclosed that in 2017, *Wikipedia* had to secure Garfield's information page after a 60-hour editing war that was stimulated by an interview that Garfield's creator, Jim Davis, had given to Mental Floss, an American digital, print, and e-commerce media company.

While the associated article, entitled "20 things you might not know about Garfield" was first printed in 2015, it had begun trending on the online news and social networking service, Twitter, in February 2017, for one specific reason. In the interview, Davis provided insight into the

© The Author(s) 2018
E. Cashmore et al., *Screen Society*, https://doi.org/10.1007/978-3-319-68164-1_8

many dimensions of Garfield's fame, including details of Garfield's 17 million fans on Facebook, a full-length stage musical dedicated to the character, and a syndication in more than 2500 newspapers and journals. But despite the various nuggets of information offered by Davis (and according to our Screenager from the US) it was the following line that captured the imagination of the general population: "Garfield is very universal…not being male or female … it gives me a lot more latitude for humour."

Drawing on this one statement, the cartoon cat soon served as a focal point to discuss and debate wider conceptions of gender through the medium of the computer screen. For instance, social satirist, Virgil Texas tweeted: "Garfield has no gender" before going on to alter Garfield's Wikipedia profile, changing gender from male to none. Less than one hour after Texas's change, Garfield's gender reverted back to male and one minute later an editor in the Philippines made Garfield genderless again. According to the *Washington Post*, the war of words continued and escalated into a 'fact' finding/conceptual debate with one editor pointing out that Garfield is constantly referred to using male pronouns (a reminder that up until the late 1960s gender was largely used as a grammatical tool, not necessarily to demarcate social identity), and other aficionados taking to back-issues of associated magazines that might reveal gender details, one suggesting: "Garfield may have been male in 1981, but not now." The *New Zealand Herald* summed up the online debate in the following way: "The character's listed gender vacillated back and forth indeterminately like a cartoon version of Schrodinger's cat: male one minute; not the next."

The analogy made to Schrodinger's cat is tenuous, but useful in some respects when discussing the concept of gender. To explain, Austrian physicist Erwin Schrodinger set out to illustrate flaws of the Copenhagen interpretation of quantum mechanics in which it argues that a particle exists in all states until it is observed. In order to dispute such claims, Schrodinger designed a hypothetical experiment in 1935 where he asks us to picture the following scene: a cat is placed into a sealed box alongside a radioactive sample, a Geiger counter and a bottle of poison. If the Geiger counter detects that the radioactive material has decayed, it would trigger the smashing of the bottle of poison and the cat would be killed. His point was this: it is not possible to uphold the logic of the Copenhagen interpretation of quantum mechanics in which the cat is both alive and dead until the box is opened. Instead, Schrodinger insists that the cat is either dead or alive, there is no in-between.

Of course, Schrodinger is speaking on a specific issue as a physical scientist, and as such, he needs only to consider physical and not social entities. But suppose for one moment that we were to apply a similar philosophy to the social scientific treatment of gender. All of a sudden, the outcome is not so straight forward. Why? Those involved in the Garfield gender dispute of 2017, supported by numerous *Screen Society* participants would testify that binary conceptions of gender and gender roles tend not to be universally accepted. After all, gender is thought by some to be a liquefied concept in which agents can be anything that they want to be, at any time.

For example, some Screenagers indicate that internet technology has helped agents to open the minds of world citizens beyond dualistic possibilities. One female Screenager in her forties from Kentucky, US, reminds us: "With the rise of forums like *Stay at Home Dads*, there's a visual representation of changing gender roles, and the net helps to legitimise it". Another female, in her twenties from Stockton writes: "the "This Girl Can" video that was posted on YouTube and all over Facebook empowers young women to take up traditional male sports such as boxing and football", whilst one man in his sixties from Liverpool points out: "males can become famous on Instagram for being make-up artists which can typically be seen as a female thing."

It appears then, that gender exists in various permutations as seen through the eye of the beholder. Human beings can be categorised as male, female or nonbinary based on the viewpoint of the agent in question and moreover, there is now a clear space for people to explore, challenge and reaffirm conceptions of gender via internet communication, facilitated though the screen.

Much of the research that exists on the subject of gender and the internet, however, reinforces binary distinctions. For instance, Chin-Siang Ang in 2017 set out to uncover gender differences in internet habit strength and online communication; Amirnima Negahdari moves to explore gender differences that influence online buying in 2014; and in 2010 Bassam Hasan attempted to uncover differences in online shopping attitudes. In all cases outlined above, and in many other studies too for that matter, what's usually lacking is the in-depth opinions of those directly implicated—the agents themselves.

In this chapter, we seek to rectify this position—but first we recap on the historical associations between sex and gender in order to contextualize the experiences and views of our *Screen Society* participants.

Men Are from Mars …

Imagine that men and women are from different planets, and by a twist of fate they simultaneously construct space rockets (blue and pink, 'naturally') and head to planet earth. Both speak the same language, well kind of—but they don't understand each other fully because they are in-effect different biological species with different psychosocial profiles. This, of course, is part of the narrative from the 1992 best seller *Men Are from Mars Women Are from Venus*, written by American relationship councillor, John Gray. It sounds far-fetched, right? But actually, it's not too far removed from scholarly conceptions of gender across time. After all, as the concept of gender has evolved, it has been used as a derivative of biological sex with social behaviour explained in relation to established differences between men and women. Within the realm of functionalist sociology for instance, scholars promoted what became known as Sex Role Theory, which explained how men and women are 'naturally' suited to particular social and vocational roles; women to more expressive roles due to nurturing instinct, men to more instrumental roles due to competitive, aggressive dispositions.

In 1949 this view was challenged by the French Feminist Simone de Beauvoir who published her seminal book *The Second Sex*. In it she declares: "One is not born, one becomes a woman." She means that while sex is undoubtedly biological, gender is not. Rather, she insinuates that gender is learned, practiced, performed and felt, and this is a view shared by many contemporary scholars.

It would seem, then, that de Beauvoir had turned popular thinking on its head by highlighting that the subordination of women and girls across history is not due to natural order, but rather to the social construction and recurrent practice of gender. Other important publications too, such as Margaret Mead's *Male & Female: A study of the Sexes in a Changing World*, in 1949, and Ann Oakley's *Sex Gender and Society* published in 1972 reaffirmed this point by drawing on the discipline of anthropology in order to demonstrate that there are variations in the way that different cultures define gender and gender roles. They argue that because there is no universal practice of gender, it must be constructed in order that men hold hegemonic forms of gendered existence across cultures and across time.

This narrative however, was only partially accepted by influential scholar, Judith Butler, who in her classic book *Gender Trouble*, published in 1990, praised feminist writing for drawing political attention to social inequality, but simultaneously pointed out that there had been an unintended consequence. Feminism, up to this point had involved an identity based theory which unintentionally served to reinforce the binary gender hierarchy that it fundamentally opposed. In other words, women and men had been categorised as distinct from one another, but at the same time, each of the categories (women/men) were defined as a coherent group that shared common characteristics and interests. This, Butler asserts, was a mistake.

To address this issue, Butler moved focus away from identity theory in order to centralize the performative nature of gender. Gender performances, she explained are the result of imitation of actions and rhetoric consumed through everyday human communications and through mediated discourse. The end result being that processes of social imitation of the type described above, would often lead, though not always, to cultural performances of masculinity and femininity.

But far from arguing for a reductionist account of gender that would situate agents as cultural dupes, Butler explains that gender performance involves an open process of repetition which always invites the possibility of change. Thus, people are not to be categorised as masculine, feminine, gay or any other one identity. In an era of globalisation (but writing before the inauguration of the World Wide Web, in its popular form at least) Butler contends that it is the logic of performance, rather than traditional notions of identity, that is most appropriate for grasping how gender is played out in late modern life.

So how do scholarly approaches to gender resonate with the reality experienced by *Screen Society* participants as they consider the impact of the World Wide Web? Does masculine domination exist on the internet or has the web changed public perceptions of gender? Well, of the 1400+ contributors to phase 1 of the *Screen Society* project, 40% agreed that the internet has helped to challenge traditional gender roles, while 60% said that they thought it had made no difference at all. Below we explore the dominant issues raised by Screenagers, beginning with participant concerns for the nature of online communications.

To Change a Lightbulb?

How many male chauvinists does it take to change a lightbulb? The answer, according to the *Jokes4US* website is: "None. Let her do the dishes in the dark." But it's only a joke, so there's no need to make a big deal of it, right? Wrong!

"It is important to note that expressed sexism doesn't have to be characterised by hostility", explains Jesse Fox and colleagues from the School of Communication at Ohio State University in 2015 (p. 436). After all, one of the defences of sexist comments is that they are intended as a joke, or are described as a light-hearted form of dark humour, otherwise known as banter. Conversely there are those, such as Martin and colleagues who, when investigating humour and psychological wellbeing in 2003, argued that banter can demonstrate a level of approachability and friendliness that enables people to criticise without ever producing negative interpersonal effects. But contrast this with the thoughts of *Screen Society* participants who were considering the impact that the internet has made on gender relations.

A woman from York in her thirties provides a damming assessment of the internet, stating: "it's almost like it's set back in the 1970s again where sexist jokes and belittling women was standard practice and women just had to put up with it or risk more ridicule." This view was upheld by numerous others including a female in her twenties from Glasgow who explained: "the internet is a bit like the Wild West for women … it can be intimidating."

For these participants and many more like them, the internet is simply a vehicle for the cultural repetition of gender inequities, or as a female respondent from the US claimed: "the internet and its trolls uphold the misogynistic cultures of the world" (for more on internet trolls, see Chapter 7). Moreover, misogynistic abuse directed towards women was felt across cultures with participants from across Europe, America, Canada, Australia and Africa registering similar concerns. As a typical example, a participant in her fifties from Ontario makes the following point: "The level of abuse directed at females on the internet and the content it contains shows that a lot of men are quite antiquated and also disturbed."

What our Screenagers seem to be experiencing, according to David Muggleton (reported in his book *Inside Sub-cultures*, published in 2000) is "the effects of core membership" which tends to privilege masculine criterion—and according to scholar Jessica Megarry, this is now a reality in large proportions of internet space. In part, Megarry's evidence is derived from a 2014 experiment in which she recorded all tweets that contained the following hashtag over the course of one month, #mencallmethings, and then she inspected the content.

Based on the level of abuse that women tweeters received, Megarry concluded that online sexist comments can impede freedom of expression and cause women to modify their own behaviours in response. She argued that while women have safe hiding places online (she's referring to spaces not yet inhabited by males, knowingly at least), the voices of women that publicly and politically challenge male social dominance are often silenced or abused. In this regard, the anonymity of the web is often cited as a cause of heightened levels of abuse towards women. However, it is important to note that the anonymity offered by the internet does not always hold negative implications for women and other marginalised groups. We will return to this later in the chapter.

All the Worlds a Stage

Act II Scene VII of *As You Like It*, the William Shakespeare play believed to have been written in 1599, begins with the following words: "All the world's a stage and all the men and women are merely players."

But was Shakespeare right? Is life simply a performance where we all understand our roles and act out accordingly? Some 360 years after the inauguration of the play, Canadian–American sociologist, Erving Goffman gave his view on this question. In his 1959 book, *The Presentation of the Self in Everyday Life*, which applies a dramaturgical approach to the process and meaning of 'mundane' interaction, Goffman argues that social interaction can be likened to a theatre, and people in everyday life to actors on a stage who play a variety of roles. Impression management is at the heart of Goffman's work, in which he explains that people, wherever possible, will present themselves in a manner that avoids embarrassment of themselves and others by playing up to the expected social role. Of course, one of the most commonly played roles in everyday life is gender.

So, for Goffman and for 60% of our *Screen Society* participants for that matter, Shakespeare was on to something! And if it was possible to transport England's greatest playwright to the year 2018, he would likely feel vindicated by the longevity of his philosophical musings—especially in relation to gender roles. Sure, it would take him a while to acclimatize to our fast moving technologically laden, globalized world—but, according to a proportion of our participants he'd soon discover that men and women remain typecast and that self-presentation and gendered role-play is alive and well. The following quote from a female participant in her thirties from London typifies the view of those supporting this position:

> I think there is a worrying trend in the definition of gender roles. More than ever I am becoming aware of the fact that women are almost always represented in a stereotypical manner presented as meek, or defined by their relationship with a man. I don't feel my experiences on the internet have made me feel any different about this.

But not all participants agree that women are typecast against their will. A man in his 20s from Edinburgh insists: "it's all out of proportion… we all play-up to our gender roles." In agreement, a middle-aged man from Essex states: "gender roles are consensual and therefore not something that should feature as complaint." He asks: "have you seen the social media photos that people are willing to share? People want to be stereotyped as masculine or feminine." Adding to this, a female participant from California explained: "Men and women promote themselves on social media in a way that they want to be seen, successful, happy, suntanned and sexy."

For French social theorist and predominant intellectual, Pierre Bourdieu (1930–2002), the findings above would go some way to help explain how gender stereotypes, and thus, *Masculine Domination* (the title of his 2001 book) continues, albeit with modifications into the *Screen Society* age. First, he notes that we all have history that has shaped us into becoming—well…us! Our values, dispositions, likes, fears, ambitions, aspirations, attitudes (amongst other effects) are inscribed upon us as we interact with loved ones, peers, institutions, and mediating influences—including our screens. In relation to gender, he explains how the amalgamation of these factors can create a gendered habitus; that is an internalised view of how men and women ought to act from one moment to the next.

When we are born we are assigned a colour of clothing, usually blue or pink. When we start school we are allocated different uniforms (e.g. trousers or a skirt). Our parents or guardians make sure that we form closer associations with same sex children, and that we partake in extracurricular activities that are 'gender appropriate'. As we grow, men and women play sports independently of one another, wear different shaped shoes, drink converse alcohol beverages, buy different colognes, button our shirts on opposite sides, use alternative washrooms in public spaces, get our hair cut at different type of establishments (hairdressers/ barbers), and are ultimately measured against different life criteria.

The list goes on, but with one common denominator. All of these differences are constructed. Writing in 2015, Raewyn Connell and Rebecca Pearse remind us, just about everything we do, think and feel relates, in some way, to our identities as gendered beings—and this was exactly Bourdieu's point too. And while the authority of the gender narrative can be questioned, challenged and changed—it tends largely to retain form across time as agents fall victims of symbolic violence and misrecognition of the implications of action.

Bourdieu means that men and women (underpinned by a gendered habitus) tend to be complicit in the maintenance of symbolic order as they misrecognise the implications of their everyday actions. This is captured below by the observations of a woman in her twenties from Loughborough who outlines the proliferation of 'typical' online gendered profiles:

> There's so many pictures online. Muscly boys and also many girls profile pictures on social media feature them wearing heavy make-up and have edited photos to look more attractive and conforming to gender stereotypes in the hope of getting a like.

Getting a 'like', as this participant infers, is the equivalent of getting a pat on the back, a well-done sticker from a teacher at school, positive recognition from fellow peers, or a certificate to identify some form of achievement. It's how we've always measured our self-worth as we exist across various social fields. Bourdieu refers to this process as the accumulation of social capital, which symbolizes power and authority and creates the unwritten rules of acceptance in different cultural fields. And in the age of the screen—it's found yet another application on the web.

Social capital involves "a continuous series of exchanges", writes Bourdieu in 1986, and the internet duly obliges. In 2016, Alex Lambert from Monash University, Australia reported briefly on the way that Facebook users watch each other as a "kind of social capital in process" (p. 2570). Lambert explains that when people become Facebook friends they implicitly grant each other access to public information that they choose to publish. Thereafter they then get the opportunity to like the content which in turn has a positive impact on the information producer. They begin to give off, what Goffman calls tie signs, by signifying to others that there's a shared relationship that is based on common values, dispositions and views on capital accumulation.

However, in the same way that it's possible to acknowledge a relationship and to reward through social media, it is also possible to punish by defriending, blocking, snubbing and ignoring. But when it comes to gender and sexual desirability more specifically, what is it that gains us capital? Simple answer: it's our bodies.

Docile (Gendered) Bodies

Back in 1993, Howard Rheingold, in the book *The Virtual Community: Homesteading on the Electrical Frontier*, argued that "people in virtual communities do just about everything that people do in real life, but leave their bodies behind" (p. 3). While this seems like a reasonable statement for its time, it is at odds with the experiences of Screenagers who have embraced the user potential of Web 2.0 and opted to take their bodies with them, albeit in digital form. After all, internet users may not be physically present in these spaces, concluded Danah Boyd and Jeffrey Heer in their 2006 assessment of networked identity, but their virtual presence takes on a different form of digital text or images which are supported by platforms such as Twitter, Facebook, and Myspace.

In relation to those platforms noted above, writing in 2010–2011, Niels Van Doorn explained that networked technologies are shaped and depend on performative practices such as the swapping of likes, kudos, photographs and general commentary—of which the overrunning theme is approval or disapproval of appearance and presentation of lifestyle. He explains how social exchanges online are "overflowing with gendered and sexualised affection" and that this is crucial to the network structure (p. 535). Moreover, as participants reveal, the surveillance of bodies online has an effect on how they think about and present themselves to others. In other words, gender is not so much a property, quality or even a description, but a performative act (i.e. a performance characterized by a cultural role).

"We are all under surveillance and this makes us conform to gender expectations," writes a woman in her fifties from Surrey. She makes the point that endless surveillance on the web when combined with gendered 'norms', mediated information, and commercial advertising, provides the conditions that can perpetuate and therefore maintain gender specific behaviours. This sentiment is one that would not be lost on scholar Alice Marwick, who when writing in 2012, explains that when we use the web, we are in effect spying on others and consistently forming judgements. She coins the phrase "online social surveillance" to explain how we use social media sites to monitor content created by others, to share information, and regulate the content of that information based on the likely perceptions of the audience. Crucially, she argues, users are motivated by social status, attention and visibility, as the following female participant in her twenties from Cumbria reaffirms: "the internet provides the platform for us to monitor ourselves and others." She continues by outlining: "I don't think it's always a good thing, because people end up putting pressure on themselves to look right."

As an extreme but common form of surveillance relating to social media, participants were keen to mention the practice of body shaming. This refers to inappropriate negative statements and attitudes relating to another person's, look, weight or size, and according to a participant in his early forties from Newcastle: "it's a common occurrence on Twitter with people passing comment on how others look." Another male in his thirties explained: "there are pressures for men and women to act in certain ways on Instagram", and a female in her twenties from Sydney said:

> most body shamer's of women are women … we live our lives through the internet, and "celebrity" culture defines what women should look like. You only need to look at the number of airbrushed selfies on Facebook and Instagram to see that's true.

Two 2016 studies (with contrasting methods) are indicative of online body surveillance as described by Screenagers. First, Seong Ok Lyu from Dongseo University, in the Republic of Korea, examined the psychological reactions of 394 domestic tourists aged between 20 and 30 on personal responses to numerous Likert scale statements such as: "I believe clothes look better on thin models", and "prior to posting my travel selfie images on social media, I use photographic filter applications to make me look better." (A Likert scale is a measurement used to represent people's attitudes towards particular subjects.)

In conclusion, the author notes how this sample of Korean female tourists spend time and effort airbrushing and editing their travel selfies before they are posted on social media webpages. Moreover, the research reveals how Body Mass Index (a value derived from the mass and height of an individual. It's used by health professionals to crudely categorise people as under, or overweight) tends to have a moderating effect on the relationship between internalization of sociocultural appearance ideals and shame (p. 193). In other words the research indicates that the social pressure of beauty standards and associated feelings of shame, appeared to be more intense under conditions of high BMI.

While valuable, one of the methodological flaws of the study above is that it didn't allow participants to expand on and explain their reactions to the set statements. Addressing this issue (albeit with a small research sample of only 24 participants) Trudy Hui Hui Chua and Leanne Chang of the National University of Singapore set out to extract meaning from teenage girl (aged 12–16) narratives in relation to photograph-based self-presentation on social networking site Instagram—which at the time was the most popular photo-sharing platform for teenagers. Participants were selected from Singapore secondary schools via a snowball sampling technique, and following the analysis of interviews, the authors concluded that teenage girls negotiate their self-presentation efforts to achieve the standards of beauty projected by their peers. Moreover, the social media tools—likes and followers were used to measure and grant peer approval of physical beauty (p. 195).

The work of French historian and philosopher, Michel Foucault is useful for decoding those findings outlined above—because for Foucault, surveillance is power. In his 1975 publication, *Discipline and Punish*, Foucault explained that to induce a person to a state of permanent visibility will create the conditions necessary to ensure that written and unwritten rules of practice are adhered to and social order is maintained. He argued that because constant surveillance (which the internet intensifies) makes every 'wrongdoing' visible, then agents become obedient citizens. When applied to gender, it is the obedience to established gendered norms that helps to produce and recreate docile bodies. This means that gender becomes transcribed on our bodies as agents follow and respond positively to normative behaviour—and of course, judge others accordingly. As the end result—each individual exercises surveillance over and against themselves, and thus, people conform to gender norms out of fear of making oneself conspicuous and therefore ripe for ridicule.

We should add that the very notion of gender norms is open to challenge: it suggests that there is typical, or standard types of social behaviour appropriate to gender. While this may have been the case, in the 1970s, when Foucault was writing, it seems a lot less tenable in times of gender neutrality and sexual fluidity. It should also be noted that not everyone obeys gendered stereotypes and the internet provides a liberating space for those not wishing to conform.

The World in the Palm of Your Hand

"Essentially, you could say, the iPhone in your pocket is a super-computer", says Professor Anthony Giddens speaking in 2015 at the Hertie School of Governance, Berlin. "It's more powerful than the biggest super-computers were 30 years ago, and there it is in your pocket! You've got the whole of human knowledge accessible to you."

Resist the temptation to Google it and just think for a moment. Giddens is right, isn't he? We do have the whole world at our fingertips. In fact, one click here or there can open up new worlds and more opportunities than we ever thought possible. From our mobile phones, computers and tablets we have access to endless information. We can communicate with people and with various robots across the globe (robots meaning: machines capable of carrying out a complex series of actions automatically). Consequently, those communications have an impact on our lives in the moment and shape our views and actions into the future.

Giddens' views in 2015 are mirrored in his book *Consequences of Modernity* written 25 years earlier, in 1992. In it, he asserts that we live in a world of high risk/high opportunity. It is high risk, he explains, because it is a world beyond control as the digital revolution catapults us all into the unknown. But it is also a world that offers immense personal opportunities.

For Giddens, reflexivity (that is the way that people reflect on what they are doing) is crucial to how society constitutes itself, but he also acknowledges that there has been a radical intensification of reflexivity in the digital age. Self-monitoring and social relations, he asserts, are now intertwined. Writing before the rise of Web 2.0 (which has intensified the capacity for reflexivity even further) he states:

> The reflexivity of modern social life consists in the fact that social practices are constantly examined and reformed in light of incoming information about those very practices, thus constitutively altering their character... Only in the era of modernity is the revision of convention radicalised to apply to all aspects of human life, including technological intervention into the real world. (p. 38)

Reflexivity is more than reflecting on our own actions. It also means reflecting on incoming information in all of its various guises. According to 40% of Screenagers, it is the exposure that they have to information from all world citizens that provides opportunities to challenge traditional conceptions of gender. One participant in his fifties from Gloucestershire believes: "reaching broader audiences can help to dismiss myths and to help individuals to see that there are different ways of being male."

The myth that this participant is referring to is the assumed presence of a singular masculinity, which according to Raewyn Connell, in the 2005 book aptly titled *Masculinities*, there is an abundance of evidence to suggest that masculinities are multiple, with many internal complexities and even contradictions. For instance, some men may hold feminine characteristics and some women may hold masculine characteristics. After all, masculinities do not speak just about men, but they are concerned with the position of men and women in the gender order within and across cultures. In other words, as social agents we are constantly constructing and reconstructing gendered identity through social exchange specific to our influences, which can be vast and varied.

As well as myth slaying: "the internet can inspire individuals to look beyond gender roles" writes a male participant in his forties from Edinburgh. He proclaims, that young people can now find real life role models (rather than celebrities) and he provides the following examples: "A young girl in India can now be in contact with a female engineer in Germany. A young boy from London can watch videos of a male ballet dancer in America."

This finding was common amongst parents and guardians who use the internet as a tool for parenting, especially tutoring daughters on personal ambitions and aspirations. One participant, in his fifties from London explains: "The internet allows me to show my daughters examples of women who have broken the traditional gender assignments and that they can set out to achieve whatever they set out to without feeling a profession/job/pastime is 'for boys only'."

Of the 40% who agreed that the internet has made a difference to conceptions of gender role, all agreed that available information/education and self-reflexion were crucial in this process. All were aware that humans are only one click away from pursuing answers to questions that they might never have had the courage to ask in 'real life'.

You're a Dog

"One of the major problems with the internet", said a *Screen Society* participant in his fifties from Holland "is that people will say things that perhaps they would refrain from in everyday personal contact." He is referring to the effects of anonymity, or the online disinhibition effect, as was coined by John Suler in 2004 and expanded upon in *Psychology of the Digital Age*, published in 2016. The premise is simple: anonymity promotes higher levels of sexism, and for Fox et al. when writing on the subject of perpetuating online sexism in 2015, it's the disinhibiting factor that can lead to more serious forms of harassment. But how?

According to Suler, it's the inability to observe others directly (meaning that non-verbal actions which typically play a regulatory role in face to face encounters cannot be considered by communicators) that removes some of an individual's concern for impression management (to borrow another term from Goffman). However, as the reader will have already spotted, the conditions described by Suler do not necessarily lead to negative implications for women who may feel repressed in day-to-day communications. After all: "On the internet, nobody knows you're a dog."

This is a quote from the now famous cartoon caption, created by Peter Steiner and published in *The New Yorker* in 1993. The cartoon features two dogs, one sitting at a computer speaking to another dog who is listening while sitting on the floor. Its aim is to acknowledge the potential pitfalls of the anonymity of the net where you never really know who you are communicating with. As you might expect, academic research has been influenced by this notion too. There are studies that discuss the role of the internet in the deception of electronic commerce (see Xiao and Benbasat 2010); professional virtual communities (see Joinson and Dietz-Uhler 2002); online reviews (see Fusilier et al. 2015); social media relationships (Hancock et al. 2007); and fake news (Conroy et al. 2015).

While such studies are disparate in content and depth of analysis, the conclusions share similarities. Deception has negative implications. But for some *Screen Society* participants it is the anonymity that exists on the web that offers an important evolutionary opportunity for repressed groups (in this instance women) to gain confidence without fear of reprisal that would follow in the physical world of face to face contact. A female participant in her forties from Washington DC explains more:

> On the internet nobody knows you're a dog. And, likewise, they don't necessarily know I'm a woman unless I choose a gendered name or avatar. I think, in some small ways, that uncertainty can help disconnect us from preconceptions about how we interact. It also provides us much more awareness and interaction with people who do not conform to gender norms than most of us would experience in our day to day lives.

Anonymity can alter experience and provide confidence to women. Research into education, for example, by Caspi Avener and colleagues in 2008 has shown how women tend to feel marginalised in classroom discussions as clear pecking orders are established in physical space. And yet, when online discussions are added to the educational environment, women were shown to feel less intimidated. After all, in asynchronous communication (taking place outside of real-time), the authors conclude, dominant characters cannot interrupt the message. In other words, the internet can provide a protective environment. This notion was copied across for numerous Screenagers including this female from Manchester: "online, gender doesn't matter. People are released from the stereotypes and discrimination of the analogue world", whilst a participant in her sixties from Birmingham reminds us: "on the internet anyone can be anything they want ... there is no need for the strict gender roles as in real life."

A Stronger Voice

Mechanical solidarity, as expressed by Emile Durkheim in 1893 is largely considered to be a thing of the past. We have explained in Chapter 1 that in its original form, mechanical solidarity is thought to be symptomatic of a more simple way of life, where communities were small scale and communications with the rest of the world were limited. As the argument goes, the rise of industrialisation, larger scale communities, a complex division of labour and global markets, have contributed to a situation where mechanical solidarity was usurped and replaced by organic solidarity, which took hold of the reins when mechanical solidarity let go. This meant that people now held interdependent relationships that were fitting of more 'advanced' societies. Large-scale communities were now bound together because of collective benefit rather than the shared principles of its members.

This might have been true for a while at least, but with the advance of internet technology and Web 2.0, more specifically (over 100 years later), it's worth taking a closer look. According to Durkheim, what matters in communities that hold mechanical solidarity at heart is shared values, common beliefs and collective consciousness—all factors used by *Screen Society* participants to explain their involvement in what they considered to be online communities. Moreover, this was deemed to be particularly important for agents that could not identify with traditional gender identities.

"The internet has given people who were once a minority a stronger voice", says a participant in their thirties from Marquette. She continued by adding: "instead of knowing one person that identifies as a different gender to the one they were given at birth, people can find others and join groups which gives them a presence." Likewise, a teenager from Connecticut explains: "I'm much less afraid of being out as a non-binary lesbian online, and have found friends that accept my gender and romantic orientation."

These findings resonate with the 1992 Anthony Giddens book, *The Transformation of Intimacy*, in which he explains that late modern life has witnessed a revolution in sexuality, largely because people are freed from the needs of reproduction. After all, sexuality is no longer practiced through pre-given roles. Instead, it is nurtured through reflexively forged relationships, which in turn, have been helped along by the revolution of World Wide Web communications.

As one of numerous examples of the ways in which sexuality can be expressed online, Michael Green and colleagues set out in 2015 to discover how the LGBT communities use the online platform, YouTube. The authors analysed the content of 100 short uploaded videos (relating to LGBT content) and revealed the following points. YouTube videos provide a platform for self-disclosure in relation to the video-blogger, and self-discovery or affirmation for the viewer. Moreover, through participation in video sharing communities, agents can improve interpersonal relations, enhance mental health and help to change social attitudes through video production.

The authors explain that YouTube is not simply a place to broadcast content, rather it is a community where people can interact around the topic of sexual identity and a shared interest in video creation (p. 704). Whilst our participants did not refer to this issue directly, on numerous occasions they did refer to the practice of personal video streaming as a way to "reach out to minorities that share similarities"; "broadcast controversial issues"; "listen to alterative views" and create situations that "bring societies outcasts out of the woodwork and into the open."

After all, when geographical proximity is taken out of the equation, those agents that feel isolated within their physical world can find a voice by locating and participating in virtual communities. For instance, a participant from New York upholds this position when talking of her parental experiences:

> Even though they were assigned female gender at birth, my child now identifies as non-binary … In my view I'm not sure that they would have come to this conclusion without finding out via the internet that there were other people who felt uncomfortable in their assigned genders.

Thus, as well as serving to provide online communities for those looking for mechanical solidarity in the global age, the internet also serves as a source of information that has the capacity to educate and therefore challenge conceptions of gender identity. In the situation outlined above, the participant describes how, as the parent of a non-binary child, the internet was an invaluable resource as she became aware of and "learned to navigate the world of more than two gender identities."

Such examples emphasise the role that reflexivity plays in practical and public life, as agents attempt to look at the circumstances that shape their social existence in order to explain new personal experiences or the experiences of others. In sum, just about every component of *Screen Society* has been opened up to scrutiny. The online world is relentlessly seeking clarification, asking questions, challenging dominant thought patterns and material thought once to harbour steadfast 'factual' information. In the end, only one consistency remains. Everything we think that we know is being liquefied by the net, and this includes our gender.

References

Albrechtslund, A. (2008). Online social networking as participatory surveillance. *First Monday, 13*(3). Available at: http://www.uic.edu/htbin/cgiwrap/bin/ojs/index.php/fm/article/view. Accessed May 2015.

Ang, C. (2017). Internet habit strength and online communication: Exploring gender differences. *Computers in Human Behaviour, 66*, 1–6.

Avener, C., Eran, C., & Kelly, S. (2008). Participation in class and in online discussions: Gender differences. *Computers and Education, 50*, 718–724.

Bourdieu, P. (1986). The forms of capital. In J. G. Richards (Ed.), *Handbook of Theory and Research for the Sociology of Education* (pp. 241–258). New York: Greenwood Press.

Bourdieu, P. (2001) [1998]. *Masculine Domination*. London: Polity Press.

Boyd, D., & Heer, J. (2006). Profiles as conversations: Networked identity performance on Friendster. In *Proceedings of the Hawaii International Conference on System Sciences (HICSS-39)*, January 4–7, 2006, Persistent Conversation Track. Kauai, HI, IEEE Computer Society.

Butler, J. (1990). *Gender Trouble: Feminism and the Subversion of Identity*. London: Routledge.

Connell, R. (2005). *Masculinities*. London: Polity Press.

Connell, R., & Pearse, R. (2015). *Gender in World Perspective* (3rd ed.). Cambridge: Blackwell.

Conroy, N., Rubin, V., & Chen, Y. (2015). Automatic deception detection: Methods for finding fake news. In *Proceedings of the 78th ASIS&T Annual Meeting: Information Science with Impact* (pp. 1–4). St. Louis, MO: American Society for Information Science.

de Beauvoir, S. (1988) [1949]. *The Second Sex*. London: Picador Classics.

Durkheim, E. (1893) [1984]. *The Division of Labor in Society*. London: Macmillan.

Foucault, M. (1975). *Discipline and Punish: The Birth of the Prison*. New York: Vintage Books.

Fox, J., Cruz, C., & Lee, J. Y. (2015). Perpetuating online sexism offfine: Anonymity, interactivity, and the effects of sexist hashtags on social media. *Computers in Human Behaviour, 52*, 436–442.

Fusilier, D. H., Montes-y-Gomez, M., Rosso, P., & Cabrera, R. G. (2015). Detecting positive and negative deceptive opinions using PU-learning. *Information Processing & Management, 51*(4), 433–443.

Giddens, A. (2015). *Into the digital age—The world in the 21st century*. Opening lecture of the academic year 2015/16 at the Hertie School of Governance, Berlin. Available at: https://www.youtube.com/watch?v=KrfDMOxRk7k.

Goffman, E. (1990) [1959]. *The Presentation of Self in Everyday Life*. London: Penguin Books.

Gray, J. (1992). *Men Are from Mars, Women Are from Venus*. London: Harper Collins.

Green, M., Bobrowicz, A., & Ang, C. S. (2015). The lesbian, gay, bisexual and transgender community online: Discussions of bullying and self-disclosure in YouTube videos. *Behaviour & Information Technology, 34*(7), 704–712.

Hancock, J., Toma, C., & Ellison, N. (2007). The truth about lying in online dating profile. In *Proceedings of the SIGCHI 2007 Conference on Human Factor in Computing Systems*, San Jose, CA, 449–452. ACM.

Hasan, B. (2010). Exploring gender differences in online shopping attitude. *Computers in Human Behavior, 26*, 597–601.

How many male chauvinists does it take to change a lightbulb? Available at: http://www.jokes4us.com/dirtyjokes/womenjokes.html.

Joinson, A., & Dietz-Uhler, B. (2002). Explanations for the perpetration of and reactions to deception in a virtual community. *Social Science Computer Review, 20*(3), 275–289.

Lambert, A. (2016). Intimacy and social capital on Facebook: Beyond the psychological perspective. *New Media & Society, 18*(11), 2559–2575.

Lyu, S. O. (2016). Travel selfies on social media as objectified self-presentation. *Tourism Management, 54*, 185–195.

Martin, R., Puhlic-Doris, P., Larson, G., Gray, J., & Weir, K. (2003). Individual differences in their use of humour and their relation to psychological well-being: Development of the humour styles questionnaire. *Journal of Research in Personality, 34*(1), 48–75.

Marwick, A. (2012). The public domain: Social surveillance in everyday life. *Surveillance and Society, 9*(4), 378–393.

Mead, M. (1962) [1949]. *Male and Female: A Study of the Sexes in a Changing World*. Harmondsworth: Penguin Books.

Megarry, J. (2014). Online incivility or sexual harassment? Conceptualising women's experiences in the digital age. *Women's Studies International Forum, 47*, 46–55.

Muggleton, D. (2000). *Inside Subculture: The Postmodern Meaning of Style*. Oxford: Berg.

Negahdari, A. (2014). A study on gender differences influencing on online buying. *Management Science Letters, 4*, 2203–2212.

New Zealand Herald. (2017, March 3). Debate prompts Garfield to clarify cat's gender. Available at: http://www.nzherald.co.nz/world/news/article.cfm?c_id=2&objectid=11811110.

Oakley, A. (1972). *Sex Gender and Society*. Melbourne: Sun Books.

Rheingold, H. (1993). *The Virtual Community: Homesteading on the Electric Frontier*. Cambridge, MA: Addison-Wesley.

Suler, J. (2004). The online disinhibition effect. *CyberPsychology and Behavior, 7*(3), 321–326.

Suler, J. (2016). *Psychology of the Digital Age: Humans Become Electric*. Cambridge: Cambridge University Press.

The Straits Times. (2017, March 6). Is Garfield male or female? Available at: http://www.straitstimes.com/lifestyle/entertainment/is-garfield-male-or-female.

The Washington Post. (2017, March 1). Garfield's a boy…right? How a cartoon cat's gender identity launched a Wikipedia war. Available at: https://www.washingtonpost.com/news/comic-riffs/wp/2017/03/01/is-garfield-a-boy-how-a-cartoon-cats-gender-identity-sparked-a-war-on-wikipedia/?utm_term=.9063099fad5a.

Van Doorn, N. (2010). The ties that bind: The networked performance of gender, sexuality, and friendship on My Space. *New Media & Society, 12*(4), 583–602.

Van Doorn, N. (2011). Digital spaces, material traces: How matter comes to matter in online performances of gender, sexuality and embodiment. *Media, Culture and Society, 33*(1), 531–547.

Xiao, B., & Benbasat, I. (2010). Product-related deception in e-commerce: A theoretical perspective. *MIS Quarterly, 35*(1), 169–195.

9

Gaming and Gambling

Buying a Thrill

> Gaming gives me such a thrill that nothing else in my life comes close to achieving. The notion of being in character and going into some fictional location and engaging with gamers from all over the world gives me such a sense of fulfilment when the task is successfully completed. (Male, late twenties, London)

> A bet is buying a thrill and being part of the sporting occasion rather than just watching as a passive neutral. Just look at the amount of sport on television all round the world and its popularity in places like Asia (particularly football). A win, if it happens, is just a bonus because it is about having an interest in a sporting event. (Male, early thirties, Kuala Lumpur)

What these two examples from our *Screen Society* participants indicate is how gaming and gambling are closely related activities. Yes, gambling involves the parting of money to be able to engage in this practice, but they both involve skill and judgement, rather than pure chance and randomness. They also involve a challenge that will test the mettle of the gamer or gambler. Neither is a restful activity: in fact, they both

© The Author(s) 2018
E. Cashmore et al., *Screen Society*, https://doi.org/10.1007/978-3-319-68164-1_9

involve adrenaline rush-type of thrills. In a sense, that's the whole point: they are both exciting and have become a significant component of the everyday practice for millions of people across the world. Therefore, to try and provide some perspective to the reasons behind this mass consumption, the focus of this chapter was to understand the role and purpose of online gaming and gambling in the everyday practice of Screenagers.

Consuming the Gaming World

The introduction of home consoles like the Atari 2600 in the 1970s was the first opportunity to engage in video game-focused entertainment in the comfort of the private home (coin-operated video games located in arcades were also present in the 1970s). As the home video game industry expanded in the 1980s, popular gaming computers including the Sinclair ZX Spectrum and Commodore 64 emerged and were widely consumed. Technological advances through the 1990s led to the emergence of Sony's PlayStation, Nintendo's 64 console and Sega's Dreamcast and created what Mark Griffiths, Mark Davies and Darren Chappell, in their 2003 article termed "pixel heroes" in the shape of Sonic the Hedgehog, the Super Mario Brothers and Lara Croft from the hugely popular Tomb Raider game. The result of all this was to highlight the changing nature of consumption and engagement across society as people now had a new form of leisure in which to devote time to without leaving the comfort of their own home. This has been further advanced with the introduction of consoles since 2010 including the Nintendo Wii, Microsoft's Xbox One, Sony's PlayStation 4 as well as increased opportunities to play online games through PCs, tablets and mobile phones in an ever growing and competitive marketplace.

One reason behind this demand is for gamers to have increasing opportunities to engage in the hugely popular Massively Multiplayer Online Role-Playing Games (MMORPGs). These allow people all over the world to create individual avatars that depict a role-playing character and subsequently play in this virtual environment at any given moment of their choosing. By their very nature, MMORPGs are designed to encourage interaction, collaboration and teamwork through the forming of groups (also known as tribes, guilds, clans or crews) to complete necessary tasks.

In a 2014 article examining online social interaction, community and game design, Zaheer Hussain and Mark Griffiths outline how online gaming has been transformed by MMORPGs as they have enabled gamers to experience different social and cultural norms and group structures. Indeed, the challenging and exciting nature of multiplayer games was an important feature for one male in his early sixties from Texas:

> In the Massively Multiplayer Online game, Lord of the Rings, I can be a member of the 'tribe' of players, and a member of a voluntary cooperative association (a Kin in Lord of the Rings) that can be (and usually is) open to everyone that accepts a minimum degree of social rules; but these Kins can also be closed, with all members actually related in some way. It can surprise some folks when they find a respected authority figure in a Massively Multiplayer Online game Kinship is actually a 'girl' - they/we often assume a leadership position is occupied by a male and they can often be mistaken in this belief.

Whilst acknowledging the importance of social rules and social relations in online gaming, this comment also challenged the assumption that gaming cultures are rife with masculine posturing and bullying. For Screenagers like this, online gaming provides a sense of gender neutrality when playing multiplayer games. (By gender neutrality, we mean the concept that informs social institutions and practices, including language, which is based on the repudiation of distinguishing between humans because of their sex or gender. It is predicated on the assumption that such differences are largely irrelevant in many spheres of social activity and, as such, should not be recognised.)

Despite the millions of consumers who play computer games across the world, the representation of them is often negative with the stereotype that it is often a problematic youth subculture activity. This was met with strong resistance amongst some of our participants, including this male in his late twenties from Edinburgh, who argued against this misconception:

> As long as parents or guardians are on hand to give advice and help and understanding then it's really a non-essential. Whether it be computer games, comic books, television, heavy metal music: every generation sneers at whatever medium that children and teenagers are using. Maybe when these so-called experts delve into the experiences of all of those different age groups engaging in online gaming they will realise that it is not that bad!

Reference to a wider demographic of online gamers than is often assumed was also raised in a 2009 study by Dmitri Williams, Mia Consalvo, Scott Caplan and Nick Yee, where it was found that women and older players tend to spend the most amount of time playing MMORPGs than younger generations. Likewise, in their 2008 article examining stereotypical gamer profiles, Dmitri Williams, Nick Yee and Scott Caplan indicate that virtual worlds do not just contain young people, but include those from a variety of ethnic backgrounds, gender, age groups and educational levels. These views were confirmed within our responses, with one in every four participants stating how they engage in some form of gaming activity online. Whilst more males than females participated in online gaming, it was not specific to a certain age category; rather online gaming was consumed by a wide range of ages all across the world.

One reason for this is the different gaming features present nowadays with Bohyun Kim's 2015 article on gamification referring to "alternate reality games" (such as Sombra and The Black Watchmen), where players make their moves in the real world (i.e. outside of a computer or game console) and interact directly with other players in the game through the internet as well as by email or phone. Moreover, the success of Pokémon Go on its release in 2016 showed how a mobile phone's Global Positioning System can locate, battle, capture and train virtual creatures that appear on the screen as if in the same location as the player holding the device. Therefore, given the widespread everyday use of the internet it should not be treated as a monolithic activity, particularly when the diverse individual nuances of consumption for millions of people across the world are considered.

Social Interaction

In their 2006 article 'Internet addiction', Laura Widyanto and Mark Griffiths refer to how labels concerning 'internet addiction', 'compulsive internet use' and 'problematic internet use' have often been used interchangeably and tend to concentrate on the psychological characteristics of individuals, such as depression, anxiety, self-esteem, loneliness and psycho-social well-being. It often concerns those individuals who it is believed experience negative outcomes from their internet use, such as having conflict with family and friends or impact upon their employment or education prospects. This was a feature of a 2017 *Guardian* article on Australian snooker player, Neil Robertson, where he claimed how an 'addiction' to the video games, Fifa, World of Warcraft and League of Legends, negatively impacted on his ability to practice and perform in professional tournaments. Yet, there are also many personal benefits to engaging in online gaming communities (in the same *Guardian* article it was reported how Italian footballer, Andrea Pirlo, successfully prepared for the final of the 2006 World Cup by playing on the PlayStation on the afternoon of the game).

Challenging the misconception that online gaming can be addictive, this female in her late forties from Ohio stated: "Some games are made to be addictive. It is my choice to play them or not. I think it depends on how much self-control a person has, whether they can stop or not." Likewise, this female in her early fifties from Iowa illustrated:

> We live in a world that allows us to make choices every day. Whether it is bad behaviour such as breaking the law, or destructive behaviour such as drug use or gambling, we have the responsibility to use good judgment and realize that bad decisions carry consequences. People are seeking instant gratification more and more, and the internet provides and encourages that.

However, this male in his late forties from Southampton outlined:

> Online and offline gaming are one of my hobbies but I flit between each depending on what I'm playing. Yet, if I go on holiday I don't play and I don't mind. I am not addicted - I do it because I enjoy it. It's all about

balance. At least with a game on the internet it's more involving and pro-active than just staring at a television and being passively fed programs with no interaction whatsoever.

The social interaction found in online gaming, such as engaging in often faceless communities, has also been labelled as one of the predictors of 'problematic internet use'. Yet, the *Screen Society* data illustrated an overwhelming sense of the importance of social interaction for those people that engage in the multiplayer gaming environment. By way of illustration, this male in his early thirties from Mansfield conceptualized the thoughts of many participants when he stated:

> Human beings are social creatures, and that is at the very core of our nature because it is often how we best learn to adapt and survive. Replacing what we gain from that social interaction with virtual interaction detaches a core part of our being from itself. That is not saying it has no place, you can obviously be social online, but it is all about balance. And at the core of it is looking at the bigger picture and maintaining a mental, emotional, spiritual, sexual etc. balance all round. How that is done and via what format is fairly irrelevant.

Participants also referred to other elements of self-development that online gaming can provide: "If a child is playing online games they interact with the game more than they ever would by watching television" said this female in her late teens from Glasgow, whilst this female in her early fifties from Ontario concurred: "I hear my children talking to friends over Skype or using in-game chat channels, so I feel that their social circle has moved online and that if they were parked in front of a television they would not have that shared community." Likewise, this female in her late forties from Sunderland added:

> My teenage son has spent years chatting with people in online communities, mostly these war games, and is possibly the happiest, most optimistic, balanced, level headed person I know. He seems to get great pleasure from these virtual interactions, and feels part of a team, yet acknowledges that these are confined to gaming and is equally happy to socialize face-to-face with his friends.

These were three of a number of responses that outlined the educational potential some online games provide. As one male in his early twenties from Rome stated: "I think video games can have positive effects on people and might even help kids become creative." Thoughts like this were in line with the 2014 findings of Jonna Koivisto and Juho Hamari who raise the transferable and computer-literacy skills that digital gaming and other forms of online social interaction can produce in individuals. (This would probably count as an example of the "gamification" referred to early i.e. the application of typical elements of game playing, including scoring points, competing against others and sticking to rules, to other areas of social activity.)

Despite Mark Griffiths, Daria Kuss, Joël Billieux and Hally Pontes' argument in their 2016 article how the "internet provides an augmented yet limited perspective of reality to users and allows them feelings of belongingness that may be psychologically compensating for the lack of social rewards in their real lives" (p. 194), many Screenagers provided experiences that challenged this view. For some participants, online gaming provides a sense of normality in their everyday practice, as suggested by this female in her late thirties from Amsterdam: "It can help depression, anxiety and stress by giving people a chance to talk to others who share the same gaming interests that they wouldn't be able to talk to otherwise." For other participants, online gaming provides a form of social interaction that is therapeutic, as outlined by this male in his early seventies from London: "My brother is disabled and housebound but for several years now, he has participated daily in online gaming with people around the world. They know lots about each other's lives and social circumstances. They have less time to be self-absorbed and depressed," whilst this male in his late thirties from Swansea added: "As someone who has suffered with depression for almost 20 years, playing video games provides an escape from the negative thoughts which characterise the illness."

People engaging in the online gaming community are often feared to be vulnerable to negative outcomes given the danger of excessive use, particularly if games comprised interactive and immersive features. Yet for many participants in our study, their experience of online games was somewhat different with some referring to how they provided an opportunity to exercise their imagination in a way that wouldn't ordinarily happen in their lives away from the screen, such as this male in his early thirties from Zurich:

> The impact of video games is over stated, particularly with online gaming making it a vastly more sociable pastime than in years gone by. The ability to abuse an anonymous Peruvian with your thick West of Scotland accent as you hunt him down at gun point from your bedroom in no way means that you will seek to buy a gun and chase the locals when you take a gap year/career break and head for Machu Picchu.

Indeed, this male in his late thirties from London concurred:

> I have played computer games all my life, used a phone for half of it, and know plenty of people who have done the same. I have never felt the need to go shoot up a street because I have done it on Call of Duty or Grand Theft Auto. Mass murders and rampages were going on long before the Atari was invented. It is about one's own personality, knowing the context of what you are doing, how it might affect other people and being able to think critically about your own habits. Personally I feel technology has enriched my life.

Although elements of the academic literature suggests that online gaming instils aggression and violence in people, some participants felt that this was often blown out of proportion, including this response from a male in his early forties from Ayr: "When I was young, horror and violent movies were said to cause the same effects. We live in world now which you can see people being beheaded or worse from the comfort of your living room."

Rather than distract people from real life relationships, activities and interactions, our findings illustrate how screens and entertainment are a basic human requirement in the increasingly interactive nature of modern society. These changes have also led to Zsolt Demetrovics and his numerous colleagues in their 2011 article to highlight:

> The popularity of the games suggests that they satisfy basic needs of people; therefore, they cannot be labelled simply in terms of good or bad. Instead, their characteristics could be examined from a motivational

perspective by exploring the needs and motives behind playing them without contemplating their beneficial or harmful nature. These motives can be regarded as energizing and determining factors of our behaviour. (p. 814)

As suggested by this male in his early fifties from Sydney: "You can argue that just about anything including alcohol, junk food and even illegal drugs have both beneficial and negative effects." What is clear from this response, and many others that could have been included is that engaging in online gaming is a social habit: they are practices of choice, not compulsions that sections of the medical and academic community lead us to believe. There is nothing accidental about so-called gaming disorder because the manufacturers of these games explicitly design them to be immersive. They want people to spend as much time playing the game as possible in order to generate further interest in others also consuming the product (such as through their Facebook, Instagram, twitter or YouTube accounts). Like a brewer who does not produce a drink to make people alcoholics, the remit of game designers is to create a product that will be consumed by as wide a range of the gaming community as possible in an increasingly competitive marketplace.

Online and Offline Friendships

In his book *On Individuality and Social Forms* (first published in English in 1971), Georg Simmel (1858–1918) referred to the creation of social structures that facilitate interaction when people act reciprocally. This leads to different forms of interaction and social exchanges and this has been expanded with the changes to both society and technology since the publication of Simmel's book. With regards to the focus of this chapter, online gaming increases the potential for people to engage in two-way or multiple other forms of interaction. So how does digital technology impact on social interaction? Addressing this point, Laura Rojas de Francisco, Jordi López-Sintas and Ercilia García-Álvarez, in their 2016 article "Social leisure in the digital age", argue: "The fact that digital technologies can be two-way enables the illusion of face-to-face social interaction, thereby facilitating the expansion of social networks and affecting social practices" (p. 260). This results in the creation of different social structures that allows individuals to interact socially in a variety of different ways should they choose to do so.

Within the labels of 'problematic internet use' and 'internet addiction' referred to earlier, the academic community and the wider media are often guilty of generalizing and negatively stereotyping those who engage in online gaming as likely to have deviant behaviour and emotional issues, including isolation and a lack of offline friendships. However, for many Screenagers, the social element of gaming is one of the strongest motivations to participate, with some highlighting how this has allowed for offline friendships to also emerge. As illustrated by this male in his early forties from London:

> Online friendships have replaced real friendships to a degree (some would view this as negative) but also opened up friendships across geographical boundaries (some would view this as positive). I now have what I consider to be close friends because of online gaming and web forums and have gone to see them in the United States for example.

Likewise, this female in her early fifties from Norwich stated:

> Many of my online friends are now 'real life' friends ... in online gaming men are discovering that women play too and this is helping expand my friendship network from purely a female one. For example, I speak to many of the people I initially met online in offline space now – both men and women.

For other participants the knowledge that their friendship circles have been expanded by having online friends was also seen as a positive experience of online gaming, including this response from a female in her late fifties from Colorado:

> I have been able to meet a lot of people on the internet through online gaming that I would never had a chance to meet in real life because of where we live and the distances and borders between us. Knowing those people has been a real positive for me.

Online gaming highlights the social dimension that technology brings and makes leisure even more social because of the ability to socialize with known and unknown people often located in different physical or temporal spaces. Through responses like those above we can draw on Robert Putnam's reference to bridging and bonding social capital from his 2000 book *Bowling Alone: Collapse and Revival of American Community* because it can be generated by making and maintaining social contacts with a larger number of people than would ordinarily be available in their offline everyday practice. Bridging social capital refers to weak social ties that allow people to feel informed and inspired by engaging with each other, whilst bonding social capital refers to strong social ties that deliver emotional support and understanding. As highlighted with the above extracts from participants, it was clear amongst the responses that for some Screenagers, online gaming provides a social component by shaping the structure of everyday practice and social interaction because it subsequently allows relationships to be created, maintained and strengthened. de Francisco et al. refer to this as digital social connectivity; a practice that is enhanced through greater opportunities for people to expand their social networks in their own leisure time by being regularly available online. Hence, rewards can be accrued not only from engaging in the specific game and gaming community, but the online friendships being built that in some ways act as if the players were chatting face-to-face in a local coffee shop.

Despite views like this, Sookeun Byun and a number of other colleagues examining internet addiction in a 2009 article, blame the opportunity for socializing on the internet as "a primary reason for the excessive amount of time people spend having real-time interactions using e-mail, discussion forums, chat rooms, and online games" (p. 203). However, for many participants the amount of time they spend in online games is largely irrelevant given the satisfaction and personal rewards they get by engaging in an online community. By way of illustration was this response from a male in his early forties from Dublin: "It is a very immersive experience. You can gain a real sense of community and "friends" in online environments that you would have no access to without technology. That has to be a very big positive." With responses like this across our own findings we can again draw on the

2016 article by de Francisco et al. who explain how "digital leisure activities have social properties which differ from those of traditional leisure activities. The social properties of digital technologies transform the meaning of leisure activities, creating interconnected leisure spaces where it is possible to be socially connected and available" (p. 258). Thus, as a leisure activity online gaming implies some form of social interaction and engagement with sometimes unknown others in the virtual space. By spending time in online communities it heightens the potential to acquire new contacts and friends, with interpersonal development likely to increase the more social influence each online community has in aiding the potential for social interaction.

Gambling

In a 2001 article on problem gambling, Philip Darbyshire, Candice Oster and Helen Carrig argue that there is a widespread assumption that gambling is "becoming an increasingly widespread and damaging social and health problem" (p. 185). Likewise, in their 2005 article examining electronic gambling machines, Nicki Dowling, David Smith and Trang Thomas state:

> There is a general view that electronic gaming is the most 'addictive' form of gambling, in that it contributes more to causing problem gambling than any other gambling activity. As such, electronic gaming machines have been referred to as the 'crack-cocaine' of gambling. (p. 33)

Gambling has a historical presence in society and remains a mainstream form of entertainment and popular culture across the world for millions of people, particularly with the availability to gamble online at the click of a button.

This mass availability to gamble led to an as yet unpublished study by two of the authors of this book on over 2500 gamblers located across the world to try and understand their everyday practice and individual motivations with regards to gambling. The findings challenge the myth of the gambling addict promoted by the medical and some of the academic profession by painting a picture of gamblers as rational decision makers who understand the odds and the technicalities of betting, rather than addicts unable to control their compulsions. For example, we found that gambling is an individual's choice, with 80% describing themselves as someone who 'likes a bet' rather than 16% who classified themselves as 'gamblers'. Seventeen percent placed at least one bet every day, 19% bet three times per week, 20% bet once or twice per week with the other 44% gambling on sports but not every week.

In his 2013 UK-based article focusing on the problem gambling addiction, Paul Gallagher writes how "new evidence reveals that the number of people in danger of becoming problem gamblers has reached nearly a million, while hard core addicts have doubled in six years to almost 500,000." But how does he know this is factually true? Indeed, Steven Stradbrooke in his 2011 article "Lazy media prefers myths to online gambling truths" criticizes the use of anecdotal reports and accepted wisdom that the likelihood of problem gambling increases now that gambling is available online. Referring to a series of studies conducted by the Canadian province of British Columbia and Harvard University, Stradbrooke illustrates how less than one percent were classified as "severe" problem gamblers. Even though the low percentage was also recognised in Steve Connor's 2014 article, he also writes how it leads "to associated problems such as debt, family breakdowns and mental health problems such as depression." Whilst we are not suggesting that these problems do not exist for some people who gamble, we agree with Stradbrooke when he states: "Bottom line, a small percentage of individuals can always be counted to abuse anything, be it alcohol, junk food, credit cards, plastic surgery, exercise."

Following Stradbrooke's analysis that the common responses surrounding the motivation to gamble were 'fun' and 'socialising' at the expense of 'winning money', we found that 34% gamble for the thrill of anticipation in being part of a sports event, 27.5% like the thrill of winning and 27.5% liked to make money. Summarising the thoughts of many participants, this male in his early thirties from London outlined his approach to gambling:

I bet on things to retain a sense of purpose of choosing to engage in such an event. For example, most people might watch a sports event and be neutral, but by betting on it you become part of the event and can gain extra satisfaction if you manage to win any money through your participation.

Likewise, this male in his early forties from Dublin stated:

Most of my bets are small, but I just do it as a form of entertainment to see if I can beat the bookie. It maintains my interest in the specific sports event I am watching.

However, other participants took it more seriously such as this response from a male in his late thirties from Southampton:

Betting is often stressful and involves a lot of research, but finding big value in a market can feel 'better' than most leisure activities. I never bet for a bit of fun, interest or out of boredom. For me there is a specific interest: to make money.

This is presumably a similar motivation to the one that drives people like professional poker player Daniel Negreanu, whose career earnings are over £25 million; or Gavin and Alex Walker, who have a happy nack of winning when they bet on football matches and make a good living from it. Or the hundreds of professional horse racing gamblers who bet for a living. People with gambling habits are not always unsuccessful: those who gamble and make it pay are typically ignored and not labelled problem gamblers.

Online sports gambling has risen in popularity to the point where our participants illustrate how it has now overtaken gambling in the bookmakers. For example, when the participants were asked what their primary method for gambling was, 48% said they bet online in private compared to 40% who continue to use the traditional method of entering a bookmakers' premises and the other 12% said they gambled in an online community. Of the online gamblers who bet in private, 68% use a computer, 27% use a smartphone and 5% use an iPad or tablet. This suggests computers and phones are already the most popular instruments of sports betting, with the likelihood that the use of smart-phones to gamble online will only increase further as more people consume this type of device.

The most common response to the reasons for this were the convenience online gambling provided by saving time through simply logging on to the relevant site or downloaded app at the click of a button as well as the opportunity to scour the market to see what gambling company was providing the best odds during the actual event that was taking place. For others, simply being alone was a factor in using their smartphone, tablet or computer, such as this response from a male in his early twenties from Birmingham:

> Betting in the privacy of your own home or work place means there is no outside influence on your bet. I don't like seeing other bets as that allows a sense of doubt to creep in. I have my own formula that I use and I am sure this is the same for many others.

Of the 40% of participants who recognised themselves as traditional gamblers in the sense of going into a bookmakers' premises, 62% said they did this because they continued to enjoy the social atmosphere of being with other people with similar interests. For some participants the actual human touch remained an important feature of their gambling practice, including this male in his early forties from Brighton:

> I prefer to physically see the exchange as it is easier to realise how much you are betting and you can limit how much you bet. Being in a book-makers allows you to share opinions and tends to act as a safety net for me because I tend to be more financially reckless if I bet alone.

So how much do the participants actually spend on gambling? 60% gamble 2% or less of their net income, 26% gamble between 3 and 14%, 4% spend 15–29%, 4% spend 30–49% and 6% stated they spend more than 50%. So, how realistic are they in winning? Over the course of a 12-month period 70% indicated that they ended up losing money, while 30% estimated that they came out on top. This is an interesting finding because it suggests rationality rarely attributed to gamblers: they gamble in the knowledge that there is a 70% chance they will lose their money, but it is in exchange for the gratifications gambling provides. They are not driven by compulsions or forces beyond their control, but by the same motives as someone who pays £8 or £10 to watch a movie at the cinema, or £50 a month subscription to a satellite television broadcaster. Thus, our findings challenge the orthodoxy on "gambling addiction" or "problem gambling," which based on the data we believe is largely mythical and a product of the medical profession's effort to reduce social behaviour to a physical condition (as outlined earlier in this chapter with reference to the literature on the negatives of online gaming). Based on the empirical evidence, we argue that gambling has become "medicalised" with an exaggeration of its negative effects in society. Rather like online gaming, it is a social activity that should be understood socially.

References

Byun, S., Ruffini, C., Mills, J., Douglas, A., Niang, M., Stepchenkova, S., et al. (2009). Internet addiction: Metasynthesis of 1996–2006 quantitative research. *CyberPsychology & Behavior, 12*(2), 203–207.

Connor, S. (2014, April 7). The gambler's fallacy explained? Misguided belief in the big win just around the corner could be down to brain activity. *Independent.* Available at: http://www.independent.co.uk/news/science/the-gamblers-fallacy-explained-misguided-belief-in-the-big-win-just-around-the-corner-could-be-down-9244642.html.

Darbyshire, P., Oster, C., & Carrig, H. (2001). Children of parent(s) who have a gambling problem: A review of the literature and commentary on research approaches. *Health and Social Care in the Community, 9*(4), 185–193.

de Francisco, L., López-Sintas, J., & García-Álvarez, E. (2016). Social leisure in the digital age. *Society and Leisure, 39*(2), 258–273.

Demetrovics, Z., Urbán, R., Nagygyörgy, K., Farkas, J., Zilahy, D., Mervó, B., et al. (2011). Why do you play? The development of the motives for online gaming questionnaire. *Behavior Research Methods, 43*(3), 814–825.

Dowling, N., Smith, D., & Thomas, T. (2005). Electronic gambling machines: Are they the crack-cocaine of gambling? *Society for the Study of Addiction, 100*(1), 33–45.

Gallagher, P. (2013, January 27). Addiction soars as online gambling hits £2bn mark. *Independent.* Available at: http://www.independent.co.uk/news/uk/home-news/addiction-soars-as-online-gambling-hits-2bn-mark-8468376.html.

Griffiths, M., Davies, M., & Chappell, D. (2003). Breaking the stereotype: The case of online gaming. *CyberPsychology & Behavior, 6*(1), 81–91.

Griffiths, M., Kuss, D., Billieux, J., & Pontes, H. (2016). The evolution of internet addiction: A global perspective. *Addictive Behaviors, 53*, 193–195.

Guardian. (2017, April 21). Neil Robertson says video game addiction damaged snooker career. Available at: https://www.theguardian.com/sport/2017/apr/21/neil-robertson-says-video-game-addiction-damaged-snooker-career.

Hussain, Z., & Griffiths, M. (2014). A qualitative analysis of online gaming: Social interaction, community, and game design. *International Journal of Cyber Behavior, Psychology and Learning, 4*(2), 41–57.

Kim, B. (2015). Gamification: Examples, definitions, and related concepts. *Library Technology Reports, 51*(2), 10–16.

Koivista, J., & Hamari, J. (2014). Demographic differences in perceived benefits from gamification. *Computers in Human Behavior, 35*, 179–188.

Putnam, R. (2000). *Bowling Alone: Collapse and Revival of American Community.* New York: Simon & Schuster.

Simmel, G. (1971). *On Individuality and Social Forms.* Chicago, USA: University of Chicago Press.

Stradbrooke, S. (2011, March 13). Lazy media prefers myths to online gambling truths. *CalvinAyre.com.* Available at: https://calvinayre.com/2011/03/13/business/lazy-media-prefers-myths-to-online-gambling-truths/.

Widyanto, L., & Griffiths, M. (2006). 'Internet addiction': A critical review. *International Journal of Mental Health Addiction, 4*(1), 31–51.

Williams, D., Yee, N., & Caplan, S. (2008). Who plays, how much, and why? Debunking the stereotypical gamer profile. *Journal of Computer-Mediated Communication, 13*(4), 993–1018.

Williams, D., Consalvo, M., Caplan, S., & Yee, N. (2009). Looking for gender: Gender roles and behaviors among online gamers. *Journal of Communication, 59*(4), 700–725.

10

Health

Doctor Knew Best

"If someone went onto *Dragons Den* to ask for investment in smartwatches back in the 1940s they'd be told to fuck off." At first sight this statement from a participant in his fifties might seem bizarre, but let's entertain it for one moment. The participant is referring to a twenty-first century television show, *Dragons Den* that celebrates the business acumen of the entrepreneur. On the show, a panel of business tycoons bid for a stake in good ideas (that is, ideas that have the potential to make money), and for entertainment value they mock those ideas thought never to catch on.

Suppose then, that this programme could be transported to the 1940s, as the participant above suggests. How do you expect that the businessmen and women on the panel would respond to the concept of twenty-first century personal health technology? For augments sake, let's imagine that the person making the pitch to the Dragons is a young Steve Jobs to ask for $10,000 investment in his latest design of 'Smart Watches' for 5% stake in his start-up company, Apple (in 1940 $10,000 was equivalent to $172,451 in 2017 [Apple Inc. was actually founded by Steve Jobs, Steve Wozniak and Ronald Wayne in 1976]).

© The Author(s) 2018
E. Cashmore et al., *Screen Society*, https://doi.org/10.1007/978-3-319-68164-1_10

Of the Smartwatch's many features, Jobs is able to explain that it has built-in GPS (but this spooky feature will deskill a nation of map readers, warn the Dragons). It's capable of recording distance and speed while users walk or run (but the Dragons don't see the significance of this. After all, the stop-watch is perfectly adequate). It has sensors that measure heart rate and it displays various metrics derived from physiological functions that are going on inside your body (but that's what the doctor is for, cry the Dragons). So, would the Smartwatch catch on, or would it be thought of as ridiculous and meaningless?

The point that our Screenager is making is that what people consider to be useful and relevant ultimately changes across time as the constitution of society is made and remade by agents when negotiating their environment—including the capabilities of technology. He explains further:

> People design technology based on concepts that seem socially palatable and 1940s Britain wouldn't have looked twice at a smart watch. First in the 1940s the doctor knew best and his authority would go unquestioned. Second, because of manual work, people were looking for a break from exercise and wouldn't have thanked you for the personal exercise guilt-trip watch.

From a scholarly point of view this insightful notion is reflected in the views of French intellectual, Pierre Bourdieu (1930–2002). In his 1990 work, *Photography: A Middle-Brow Art*, Bourdieu reminds us that technology is largely socially shaped as are its meanings and terms of use. In other words, the social function of technology is created not only by its makers but by its users, and as our participant notes, its users must be ready.

This trail of thought is also adopted by numerous other researchers, including Anne-Marie Barry and Chris Yuill in the 2017 book *Understanding the Sociology of Health*. They argue that if we take the time to step back and think a little deeper about the world around us it will become clear that technology is bound into social relations "and plays an important role in producing and reproducing those relations" (p. 286). After all, the scholars argue, technology does not exist in its own world, but it exists because of and is implicit within wider social relations.

With this in mind, the following chapter sets out to address a call proposed by scholars, Debora Lupton and Annemarie Jutel in 2015. The authors point out that research into digital health (and health apps more specifically) has largely taken an instrumental approach by focusing on content or legal issues—and yet few studies have concentrated attention on the everyday experiences of users. When making this point the authors highlight the need for future research to examine digital health from a socio-cultural perspective. Doing so, they suggest, will provide an insight into the evolving relationship between individuals and attitudes towards bodies, commercial health applications, even perhaps—the implications of health technology for doctor/patient relationships. As we agree with Lupton and Jutel, this is the approach that we intend to take here.

But before we discuss the impact that digital health technology has had on the lives of our Screenagers (and having already uncovered the history of the screen and digital technologies in Chapter 2) we take time to explore the evolution of what we call 'the health habitus' (a set of embedded but ever fluctuating dispositions, values and common thought processes) which has made it possible for personal health technologies to seamlessly integrate into the lives of many world citizens.

Health Habitus

According to Bourdieu in his 1977 book *Outline of a Theory of Practice*, within any given field (at any particular time) there tends to be a set of ideas, rules, regularities and forms of authority that are likely to be accepted and upheld in practice. Note: by *field*, Bourdieu is referring to the context in which social interaction occurs, and in this chapter, we locate action in the fields that intersect with health and fitness. He reminds us that whilst social fields might look or feel stable to agents in the moment, they are in fact continuously in flux (though perhaps only moving slightly) as people shape and react to emerging social circumstances, trends, political movements, business ideology, technological advancements, and many other issues besides.

Just like a domino effect, small alterations reverberate throughout cultural fields, nudging agents in the direction of change and gathering momentum as the message spreads. Those out of line with the domino rally are left standing (effectively becoming relics of a pre-gone age) while others keep pace with the trends of our times. One Screenager of pensionable age recognises this argument: "the internet hit me like a stream train, I felt out of my depth online but I knew that I had to stick with it or get left behind." Another in his early forties stated: "in this day and age you've got to keep your eye in with technology, the world is now online and you've got to keep up."

Keeping up with technology in a changing social world, according to Bourdieu, will impact on the constitution of one's habitus in the moment and into the future. For Bourdieu, habitus is a complex term that encompasses many things, but simply put it can be defined as a system of dispositions (that is—lasting, acquired schemes of perception, thought and action) that humans develop in response to the cultural and physical environment that they find themselves in. Simplified even further, our habitus is what makes us who we are and effects our capacity to make decisions in this way or that. It's principally influenced by previous cultural teachings, including ones learned aptitude to reflect and adapt.

In Screen Society, then, reflexivity is as much the habitual outcome of field requirements as any other disposition. In fact, on reflection it's more important. Changes to the modern environment have summoned conditions via which the 'reflexive habitus' has become increasingly common in the sense that uncertainty and change are becoming a familiar occurrence in most fields.

Consistent with this logic, participants in the *Screen Society* project have noticed a gradual shift in social attitudes towards exercise, fitness and health. For example, a male participant in his late forties said: "Society as a whole has become more concerned with image, but dresses it up with concerns about health." Another in his twenties suggests that social media use has made "people more conscious of what they look like and they attend gyms and diet more strictly than any other time I've been aware of", whilst a woman in her early forties refers to "today's generation" as overrun with self-absorption which manifests as a cultural "appetite for mirror gazing." She continues: "I'm not even sure that health is the objective of most people, it's probably looks. Health consciousness is just a by-product of our desire to look good that has crept up on us."

The Rise of Individualism

This creeping, shifting attitude is not lost on Marc Stern in his 2008 historical paper on the US fitness movement and the fitness cen-tred industry from 1960 to 2000. Drawing on various surveys con-ducted in the US, Stern points out that prior to the 1970s, only 24% of Americans regularly exercised. However, he explains that the 1970s witnessed the emergence of wholesale attitude change, instigated largely for two main reasons. First, in keeping with the idea that young men should be in physical condition to serve America physically if required, the government aired their concerns for national health. But secondly and perhaps most significantly from a social evolutionary perspective, attitude change was encouraged, promoted and marketed by an emerg-ing commercial leisure industry. We'll return to this later in the chapter.

To address the former issue, in 1960 US President John Kennedy set the cogs turning in the minds of American's when wrote an article in the magazine *Sports Illustrated* that was titled, 'The Soft American'. In it, Kennedy explained to the readership that "the age of leisure and abun-dance can destroy vigour and muscle tone as effortlessly as it can gain time." He warned:

> The television set, the movies and the myriad conveniences and distrac-tions of modern life all lure our young people away from physical activity which is the basis of fitness in youth and in later life … no matter how vigorous the leadership of government, we can only restore the physi-cal soundness of our nation only if every American is willing to assume responsibility for his own fitness and for the fitness of his children.

Presumably heeding Kennedy's warning, corporate America became the trend setter for the adoption of physical fitness, though the motives of this fitness trend would serve functions that differed from the philosophy outlined by Kennedy. First, exercise and fitness would be used as a form of conspicuous consumption in the sense that nineteenth century scholar Thorsten Veblen might understand this situation (i.e. using the guise of exercise as a form of social distinction for the professional class who 'do business' in the luxury, high-tech surroundings of health clubs); and secondly, it was used as a strategy to increase business productivity in the workplace.

This was shortly followed by (and ran simultaneously to) a more inclusive fitness movement that again held a different ideological and moral framework to that expressed by Kennedy in 1960. Whereas Kennedy's concern was to improve national fitness and to challenge the mind-set of the couch potato that was associated with rising leisure cultures and convenience technologies; the 1970s movement towards exercise, fitness and health quickly became fixated on individual aesthetic beauty and self-actualisation. This was captured by scholar, Dr. Warren Guild who commented in *Vogue* magazine (1971) how "Fitness has to do with vanity … anyone who says he doesn't want to look neat and trim. i.e. 'sexy' is a damn liar."

Whilst Dr. Guild captures the underlying essence of the nascent fitness market, it's important to acknowledge one further point. Fitness was just as much about consumption and economic philosophy as it ever was about health. Reflecting on this, one Screenager in his late thirties points out:

> Reagan and Thatcher [US President 1981-89 and UK Prime Minister 1979-90 respectively] shared the same idea in the 1980s which was to put responsibility for health and fitness onto the people for just about everything with the aim of reducing government spending and encouraging private companies to make and sell health-related products.

Here, this participant is referring to the emergence of neo-liberal political philosophy in the 1980s and the inevitable enhancement of the role of the private sector in the economy and, by implication, into the field of health and fitness. According to Umberto Eco, those with opportunities to make money from an emerging consumer health market did not need to be asked twice. He recounts that commercial leisure companies were quick to capitalise on emergent trends in youth culture, with sport and leisure branding used to sell dreams, lifestyles and aspirations of the beautiful body. After all, under neo-liberalism it became possible to create aspects of one's identity via the process of consumption and by focusing attention on looks.

Cultures of Narcissism

The rise of *The Culture of Narcissism* is perhaps best captured in the 1980 Christopher Lasch book of the same name where he sketches out a picture of American life in an age of diminishing expectations caused by a multitude of factors including Vietnam, Watergate, economic stagnation and the impending exhaustion of natural resources. In an environment of low expectations where Americans had lost faith in politicians, other leaders and even family groups, people convinced themselves (according to Lasch) that what matters is psychic self-improvement including the desire to get in touch with their feelings, overcome the fear of pleasure, and expand ones consciousness of health and personal growth.

As well as gracing the pages of academic publishers, Lasch's arguments struck a chord with the White House too. Most notably *The Culture of Narcissism* is thought to have influenced a speech delivered in 1979 by then President Jimmy Carter, warning of the perils of self-worship. Known as the 'National Malaise Speech' Carter said: "in a nation that was proud of hard work, strong families, close knit communities, and our faith in God, too many of us now tend to worship self-indulgence and consumption." His intention was clear: to attempt to steer Americans away from a life of self-obsession.

But like King Canute who, back in the eleventh century, so aptly demonstrated that no ruler can hold back inevitable flows (in his case, England's North Sea), Carter too proved himself ineffectual at reversing the rising tides of change. Self-indulgence, it seemed, was here to stay. It had already become a feature of the modern habitus, not only in the US, but in Britain too.

In a 1979 article entitled 'Has the ego trip gone too far?', *a* correspondent for the *Guardian* newspaper wrote about the "personal growth movement" in Britain. His argument was this: Britain's are part of a narcissistic movement, which, in the Freudian sense means "you become an object of your own interest and seek self-gratification." In the same vein that Tom Wolf had labelled 1970s America the 'into me' decade, the correspondent argued that Britain too was becoming "absorbed in the quest for self-gratification." This was typified, he suggested, by a growing desire for laypeople to sign up to personal shrinks, life coaches, therapists, fitness gurus and the demand for cosmetic surgery.

A proliferation of additional articles in the British Press signifies the scope and intensity of this movement as it entered into public consciousness. For instance, a brief inspection of *The Observer/Guardian* newspaper archives reveals articles such as: 'Slender success or is it just fat chance?' by Judy Mazel in 1982; 'Another day another diet' by Liz Jobey in 1983; 'A Healthy diet of yin and yang' by Christina Probert Jones in 1989; 'The Guaranteed truth free diet' by James Erlichman in 1992; and 'Fat is a pessimist's issue' by Gary Younge in 1999 (to list only a few related articles). These focused on the latest science-based evidence (intertwined with anecdotal rubbish) to 'educate' readers on nutrients and commercial supplements, all the while keeping them gripped with information on how to stay thin, bulk up with muscle, or lose fat.

Healthy lifestyles were also narrated through the appearance and commercial success of the high-tech gym/health club. Articles such as: 'Fitness and wealth' by George Parker-Jervis informed the readership that the fitness industry was worth £3.1 billion in 1988, "with spending on par with electricity, furniture and more than any other leisure sector including audio or DIY"; 'Gym and Tonic' by John McVicar in 1993 writes of the demand for health clubs; Cosmo Landesman in a 1995 article titled 'THE BLOKE', explains the effects of the health movement on men, with gyms and beauty products equating to "the final step in man's evolution from primate to peacock"; and Lisa Buckingham in the 1998 article 'Profit from the body beautiful' encourages the readership to join the two million Britain's that spend £1 billion in the fitness business.

As the reporters suggest, new gymnasiums or health clubs (as they were increasingly called) are businesses that attract customers by selling a fitness experience that encompasses specialist equipment, latest technologies and industry experts that are willing to share trade secrets. All of this is designed to enable people to invest in the most important person in their world—THEMSELVES!

In a more recent example, journalist Dominic Utton (writing for the *Telegraph* in 2014) describes contemporary gym goers as, "drunk on self-love and paying handsomely for it." He goes on to argue: "in a culture where appearance is more important than substance, the gym becomes a place of worship. And what people are worshipping is themselves", not only at the gym but when they get home too—because, he explains, "taking a picture of yourself working out is now as much part of going to the gym as actually working out." The typical workout, according to Utton, looks like this:

> Work up a sweat, gaze at yourself in the mirror. Take a photo of yourself to show the world. Find a similarly hot chick with whom you can validate your opinion of your own beauty then hit the showers … when things really become unbearable.

Thus, drawing on evidence derived from both Screenagers and newsprint archives, the point we are making is this: the health habitus that we recognise today is an extension of a social narrative that has been created by agents since the 1970s. Political, economic, cultural and technological advances have combined to influence individual action that is centred on self-absorption. Likewise, individual action has influenced socially acceptable ideas about health, exercise and fitness—where looking after one's health is no longer enough.

In Screen Society, the good citizen must be seen to be healthy and youthful, and if this can be achieved (not only in physical appearance, but also e-photos that saturate cyberspace) it brings rewards. In the physical world, a compliment on aesthetic look equates to an example of social capital gain; on social media, a like, a retweet, or a positive comment is all that matters.

In effect, Screen Society encourages agents to be fixated on three things: new technology, screens, and themselves. All feature predominately in the latest advancement of gadgets that are designed to aid the individual in search of their health needs. Here we are referring to the personal health app, a form of software designed to run on smartphones and other mobile devises and it is to this that we now turn.

Personal Health Apps

"Every day it seems as though a company releases a new wristband that tracks physical activity to monitor glucose levels or some other tiny, shiny smartphone-enabled medical doohickey", writes Roger Collier in his 2014 *CMAJ* commentary paper, 'rapid growth forecast for the digital health sector'. As he explains it, health apps generally fall into three categories: (1) fitness trackers to count steps and calories burned etc.; (2) real time monitoring to track vital signs in order to assist with chronic conditions; and (3) sensor-based systems to improve safety for seniors living at home.

It is largely accepted that each of these health-related categories is set to grow exponentially; and for researchers Nikhil Balram and colleagues, this is just the tip of the iceberg. In a paper published in 2015 they project that 'The Infinite Network' (TIN) (that is what the authors perceive will be the next stage of development from Web 2.0, in which all things and all people will be connected to the internet at all times) will help individuals to make the right choice at the right time on health-related issues. In effect, this trail of thought suggests that in the future, every human being will have the benefit of robotic personal advisors.

A future guided by robots might sound weird, but to an extent this happens already. For instance, a Screenager in her twenties states: "the internet is my 'go to' on health issues." Likewise, a participant in her late fifties explains: "I rely on apps to keep an eye on what I'm eating." "Having a family history of hypertension, I use an app to measure blood pressure", writes a similarly aged man with a self-proclaimed "stressful job." With this, it is worth reiterating the following point: the internet is already embedded in our psyche as a 'go to' tool to help us cope with our current lifestyles.

In keeping with Collier who reports that the most developed niche of the health gadget market is the fitness tracker, the majority of apps used by our participants, although varied, were fitness related. But whilst the plethora of apps available for use might suggest a complicated and multifaceted picture of the desire for and use of various health applications (given that apps cater for individual use), the reasons for use seem to be a little more uniform. For instance, common consensus from respondents, typified by the contribution of a male in his early forties indicates that "apps allow individuals to monitor their own physical and health related data as a tool to support fitness and healthy lifestyles." A similar aged female shared similar thoughts: "increased awareness of health-related issues allows people to take more individual responsibility for health through education and motivation." The following Screenager in his early forties put it in more simple terms: "fat people walking more because of apps. What's not to like? Less pressure on NHS."

Implicit in those statements above is an acceptance of a term that is attributed to Robert Crawford in his 1980 article 'Healthism and the medicalization of everyday life'. As the title of the article suggests, the term is 'healthism', and as Crawford explains, it describes an emerging sense of health consciousness that occurred throughout the 1970 and 1980s and is now firmly planted into the lives of Screenagers. As we can see from the responses above, healthism is (for a number participants) a general state of mind, an attitude and disposition that is ingrained on one's health habitus in the sense that it is acted out 'naturally' on and through the body. (Crawford's full definition of healthism is: "The preoccupation with personal health as a primary—often *the* primary—focus for the definition and achievement of well-being; a goal which is to be attained primarily through the modification of life styles.")

Kathleen LeBesco discusses this very topic in a 2011 article which details the moral perils of fatness. She argues that where public health strategies were once focused on hygiene issues, today it appears that the behaviour and appearance of bodies is far more important. Having a healthy body, she explains, has come to signify the morally worthy citizen, and this is a position supported by Nike Ayo in a 2012 paper discussing the subject of health promotion in a neo-liberal climate.

The point that both authors make is this: healthism is a habitual way of thinking that (whilst it appears to be driven by individuals) is complementary to and is encouraged by those holding neo-liberal political agendas. After all, by looking after their health, consumer citizens become partners in the governance of their own affairs and bodies. But more than this, there is a desire amongst some Screenagers, to raise one's profile as a health conscious citizen. For example, one participant in her early fifties explains that social capital can be raised through the consumption of goods and services, including: "buying organic foods, sports clothing, pills, powders, potions, energy drinks, diet coke, diet supplements—all carry the message—"I'm looking after my health"". She continues: "Health apps are just a new version of this process, but they can be more fun."

Fun, Fitness and Surveillance

The fun and desirable features of fitness apps, according to some Screenagers, tended to be in keeping with an apps ability to observe, record, analyse and monitor behaviours and to incite light-hearted, friendly competition with self or others. For instance, a male in his late twenties illustrated: "it's all about the metrics for me. I'm a stats man and I like to see improvements, that's the only way it keeps me motivated and disciplined." A woman in her early twenties who uses a data sharing app said "the social support is good. Sharing my data with friends is fun and we inspire each other to do more", whilst this male in his forties agrees: "by fostering social connections between amateur athletes, fitness apps have been beneficial in encouraging people to take a healthier outlook on their lifestyle." Sharing data was an important motivator for many others too. One Screenager in his early thirties explains: "I know at some point I will end up showing my stats to my mates in the pub. That's a competition that no one likes losing."

Two of these points are worth noting in relation to the commonality of participant responses. First, what appears as fun can also encourage discipline. In a 2014 special edition of the journal, *Societies*, Deborah Lupton argued that apps tend to work as disciplinary tools that create docile bodies and incite desire in the autonomous individual to follow health messages. Secondly, it appears that for some, the popularity of fitness apps is based on the concept of lateral surveillance, a term first coined by Mark Andrejevic in the 2005 essay 'The work of watching one another'.

As Andrejvic describes it, lateral surveillance relates to the process by which individuals showcase their achievements in order to seek social approval whilst simultaneously spying on peers and judging their actions accordingly. For Screenagers, social media provides the perfect platform to engage in a spot of lateral surveillance, and in this setting, fitness apps can be used to impress or to evaluate the worth of others. One participant in his thirties explains:

> I upload my latest run via the phone and it makes it available for others to see and comment on. You get likes from others that view your run and the app has a list of best times for different segments of the run so you can see how you compare to others and get ideas about where to run next.

According to Tony Rees, in his 2017 Ph.D. thesis which explores the lives of racing cyclists and features their use of the cycling app Strava, it's the balance between accurate information relating to personal health goals and the creation of an enjoyable app experience that fitness app users desire. He explains how digital apps tend to have inbuilt 'gamification' elements which heighten excitement on behalf of the user. Simplistically put, gamification relates to the application of playful context to typically non-playful situations, such as physical training (see Chapter 9 for more on gaming).

During his research (which spanned four years), Rees witnessed what he calls the 'Strava effect', where a largely physical community of training athletes began to embrace online training features, as cyclists in the observed group succumbed to the lure of digital technology. Strava became a training aid that was to be relied upon for its accuracy of information e.g. for the digital recording and analysis of training routes, power output, distances and speeds. In addition, the social media element took off too, with riders seeking new forms of social capital gained through online interaction and the peer to peer surveillance of training rides.

At less specialised levels of exercise, similar experiences were expressed by Screenagers that hold more modest aspirations for personal fitness. For example, one participant in her late twenties explained:

> I've found my health watch and its app to be inspirational. I didn't know how little I was doing every day because I'd never recorded it before. I didn't know how much I'd eaten for the same reason. It's the recording and sharing information with friends that makes losing weight enjoyable for me.

In sum, digital technology has made Screenagers think differently about health matters, and this in turn, has implications for their relationships with health professionals.

Smart Patients

"They used to say the doctor is always right but that was before computers", writes a Screenager in her early fifties. "The internet has made everyone an expert because all the information about any aliment is there for you. We are in an age where the computer is your personal health mentor." Another participant in his early fifties concurs:

> My kids just google anything they're not sure of and find the answer immediately. Their attitude to health is much more proactive than mine ever was. We were pretty much in the dark and relied on professionals who kept us at arm's reach.

The synopsis of an evolving triangular relationship between health professionals, patients and computers is not only common to *Screen Society* participants, but it is also recognised by scholars such as Balram Nikhil, Tosic and Harsha in their 2016 explanation of digital health in the age of the infinite network. They write of the Smartphone as a super-computing, super-communicating and super-sensory platform that can provide insight into the health of a person, enable better diagnostics, more effective treatment and proactivity for preventative care.

Yang and colleagues in 2012, Thomas and Bond in 2014 and Lowe, Fraser and Souza-Monteiro in 2015 share the same sentiment. That is to say, digital technologies are encouraging proactive health at the individual level by enhancing mindfulness and slowly shifting one's health related habitus from one of reliance on professionals into a cold state of reasoning based on e-information that is gathered outside of general practitioner (GP) surgeries. Perhaps most aptly, Leslie Robinson when writing for the *Journal of Medical Radiation Sciences* in 2003, coins the phrase 'smart patients' to describe a new breed of well prepared and knowledgeable agents that are reflexive, equipped to research and share knowledge as well as listen to it.

This is typified in the logic displayed by the following Screenager in his early fifties when reflecting on his own medical condition and that of his late father:

> We know more about so many conditions - physical and mental - and our food information is way better than it was thirty years ago because of the net. I have type 2 diabetes and am fit because I am informed. My father died in 1974 at the same age. Maybe if he had been born when I was, he might also have found he was diabetic and been able to do something to prevent his own death.

Like this participant, many Screenagers were well versed on all medical conditions that affect them directly and they informed us that a visit to the doctors now extends the physical appointment as patients turn to the internet for additional guidance. In fact, some had the following general message for GPs: "Please don't treat us like children and withhold information or talk in code, cos we'll check for ourselves on the internet and we get frustrated with poor information" wrote one participant in her late thirties. Other Screenagers warned practitioners that 'smart patients' are less likely than previous generations to accept the word of a professional without getting a second opinion from Dr. Google (a term that was used by multiple participants). For example, one Screenager in her late sixties writes:

> In the last weeks I was told a blood test for CA 125 was being done but they offered no other information. When I came home I looked it up and discovered that they were checking for cancer. They never mentioned that

at the appointment. It's made me sceptical and I'll be double checking anything I'm told from now on. [Note: a quick Google search reveals that CA125 is a test that may be used to look for early signs of ovarian cancer in women]

In an era of declining trust in expert authority, social theorists are beginning to explore the relationship between lay and expert knowledge. For instance, in his 1990 book *Consequences of Modernity*, Anthony Giddens writes of the increasing scrutiny of expert opinion by lay people who go on to make pragmatic calculations based on their own research. As reported above, this was true of our Screenagers too, many of whom believe that the power balance between the health professional and patient has shifted (or is in the process of shifting) because of: available information in the screen age, the rise of computer literacy, and the support that patients can muster from online communities of agents who have experienced or are experiencing the same medical condition.

As well as emotional support, scholars Devon Johnson and Ben Lowe, writing in 2015, have explained that online communities can share practical knowledge too. Patients can enquire between themselves what treatments ought to be available, discuss the facilities on offer, or, as in the following example (provided by a female participant in her early fifties) receive advice about how to jump the waiting list: "I'd been on the NHS waiting list for a knee operation and got talking to someone on a forum. One lady told me to book a private consultation to jump the cue. It worked, and I was seen quickly on the NHS."

In spite of the thoughts of some Screenagers, Deborah Lupton and Annemarie Jutel urge research theorists not to exaggerate claims of perceived authority shifts within the practitioner/patient relationship. They remind us that computers can make mistakes too, and for that reason all apps tend to come with the following caveat. If concerned, seek medical advice from 'real' experts.

So, the power balance might be safe for the time being, but Screenagers are right to point out that the personal dynamics of the relationship have altered course. For instance, in the past, people attended GP surgeries because they had symptoms and wanted to know if those symptoms were underpinning a definitive medical condition, and in turn, what treatment was required to combat the known condition.

Now patients can research symptoms online before booking an appointment at the practice. In such situations it is conceivable to suggest that the role of the doctor changes from one of information provider to a facilitator who aids with the digestion of information already collated. As one participant in his late forties illustrated:

> Gone are the days when you went to the quacks, were told that you had a condition that you couldn't even pronounce let alone spell, and were sent home with a pat on the head and a drug that might or might not help. Now you can look up your diagnosis, see what medication regimes and care pathways are available, and hopefully have a much more informed discussion with your healthcare provider.

What Screenagers like this are calling for is a model of medical care that is based on co-production. The philosophy behind co-production is this: in an overly medicalised field, health professionals ought to recognise service users in the design, implementation and decision making processes that dictate service delivery. Users should be considered as experts in their own circumstances, and unless proven otherwise, capable of making decisions as responsible citizens. This strategy appears to be embedded into the lives of many Screenagers who monitor their own health and the health of loved ones with the use of digital technology. A cross-section of experiences include:

> *Multiple sclerosis*: "My mother suffers from MS but she stays in touch with her grandchildren, through screens"; *Alzheimer's*: "For my mother in law we were able to monitor her health, mobility and safety (via camera hook ups and sensor based system) which allowed us to care for her for 8+ years at home, rather than in residential care"; *Cardiovascular* defibrillator: "My brother-in-law has just been fitted with a defibrillator inside his skin and computers monitor whether it goes off. He gets notifications on his phone. Incredible"; *Parkinson's*: "I have Parkinson's … the internet provides great detail about my condition and helpful advice"; Gall stones: "I used the internet to diagnose stomach pain which my GP said was indigestion but turned out to be Gall Stones … I went to A&E and

they confirmed my diagnosis and operated immediately"; Dialysis: "I am a dialysis patient and the internet passes time on 4-hour sessions." *Skin cancer*: "Apps are used to take high definition photographs of moles and look for signs of melanoma."

Whilst this represents only a small section of the usage of digital health technology, it's important to acknowledge that not all Screenagers advocate the use of the internet for health related purposes, nor did they all see the internet as beneficial for public health more generally. Below we give space to the cyber cynics.

Cyber Cynics and Informaniacs

The old adage "knowledge is power" (often attributed to sixteenth century philosopher Francis Bacon and used on multiple occasions by our Screenagers) assumes that insight can provide a stable base from which to question propositions or authoritative diktat in order to make informed decisions. The philosophy seems logical, but it isn't always true according to scholar Leslie Robinson who asked the following question in 2003: "Is digital technology empowering patients?"

In her answer, she acknowledges that talking to others through internet groups and watching podcasts and videos that provide vibrant and dynamic information will arguably better prepare patients for a forthcoming medical experience than it would by reading static information from leaflets and pamphlets. However, Robinson (and some of our Screenagers too) share the concern that too much information (gathered in a haphazard manner online) is difficult for the layperson to process.

Gunnar Trommer agrees, and when writing in 2015 about the gap between promise and reality for digital health, warns: "too much information is worse than not enough information since it numbs us to what is truly relevant" (p. 183). Far from being empowering to the individual, Trommer suggests that unfiltered information can hold negative consequences for health systems as they become awash with confused and anxious patients. One participant in her twenties has a word to describe the confused and anxious patients that Trommer is referring to: INFORMANIACS.

According to this and other participants too, 'informaniacs' are the hypochondriacs of the Screen Society age who, according to one guy in his early twenties "spend their time combing the net for signs of personal illness", or they "simply misread information to end up fearing the worst", states a participant in his early fifties. For Scott Lash, this type of behaviour is symptomatic of a broader process that he calls 'informationalization'. In his 2002 book *Critique of Information*, he explains that informationalization is unique to societies where the speed of information flows erodes the space needed for critical reflection. This includes Screenagers that have become dependent on the internet for the acquisition of personal knowledge.

The problem, as Lash sees it, is that genuine knowledge frequently becomes diluted, decontextualized, ephemeral and ultimately lost or contaminated with untruths and as such, it can be rendered meaningless. After all, information can be added to the web by anyone at any time and it can be redistributed, edited and rehashed throughout cyberspace via simple posts, tweets, blogs, or any other social media publishing form without question or quality control. The upshot is that agents are presented with a landscape of ever-changing information with unclear origins, content and value. Of course, in the context of health, misinformation can be dangerous.

Michael Hardy, writing in the 2013 book *Key Concepts in Medical Sociology*, floats the idea that danger to health can extend beyond the more obvious risks such as: incorrect self-diagnosis, self-medication, or following a fitness regime that is too advanced for a person with underlying health conditions. Rather, he reminds us that social media and associated 'self-help' groups that enable people to discuss information anonymously across the globe aren't always good for us. He writes of online groups such as 'Proana' groups that provide information to those wishing to lose weight to a degree that may result in clinically dangerous outcomes, not to mention the dark web and its capacity to aid steroid users that are prepared to do anything to build muscle in spite of long term health risks.

Whilst our participants did not make direct reference to such risks, a small number spoke about the capacity that the web and health apps have to reignite previously diagnosed mental health conditions.

For instance, when describing personal experiences with fitness apps, a woman in her late twenties explained how she had "an issue with a food tracking app that slightly brought back an eating disorder." She explained how the app encouraged her to "severely control" her intake of calories, and points out: "if I hadn't realised when I did I could have become obsessive and compulsive regarding it. There were no fail safes in the software to protect people from this."

Snake Oil Salesmen

As well as failing to protect those looking for a quick dieting fix or those with histories of psychological body dysmorphic issues, other Screenagers were concerned about the commercial sales of digital health related products that potentially draw on, rather than fix human vulnerabilities. As one participant in his early fifties puts it, there's a conflict of interest within a commercial company that sets out to fix health related problems for individuals because "they rely on insecurities in people who they want to buy the product", whilst a woman in her late thirties worries that fitness apps are "nothing more than a marketing exercise to extract payment from a gullible public". Another agrees that "all these fitness apps are doing is lining the pockets of companies by fuelling already sedentary lifestyles and providing a false sense of wellbeing."

Sceptics of digital health technologies (whilst in the minority of surveyed Screenagers) share the doubts of some academic researchers which were reported in a 2017 *Telegraph* newspaper article, written by Sarah Knapton, entitled 'The 10,000 steps a day myth: how fitness apps can do more harm than good'. The article heeds a warning delivered by Dr. Greg Hager, an expert in computer science at John Hopkins University, who told delegates at the American Association for the Advancement of Science annual meeting in Boston, that the 10,000 steps doctrine was based on just one study of Japanese men dating back to the 1960s.

For Hager, 10,000 steps is an arbitrary figure that is built into fitness apps with very little evidence behind it. Dr. Steve Flatt of the University of Liverpool takes this a step further to suggest digital medical apps (some of which GPs are encouraged to recommend to their patients) lack scientific scrutiny and he likens the field of digital health apps to the snake oil salesmen of the 1860s, that is, rogue traders who knowingly sold fraudulent goods to gullible customers.

So, for some scholars and for some Screenagers too, it's the conflict of interest that exists between the pursuit of technological health aids (based on sound scientific/ethical principles) and the pursuit of profits which causes most concern about digital health futures. Screenagers were aware of some techniques that are used to generate income in the field of digital health, including the practice of selling on user information to third parties, who in turn, seek to profit from the collection of big data.

The big data economy is a symptom of our Screen Society and as Prainsack suggests in a 2014 paper, it's part of a lucrative commercial market that is promoted in the interest of manufactures who sell technologies, or to pharmaceutical companies who, in turn, market goods to agents that are likely to be interested in specific products. Big data takes little effort for companies to assemble because its collection is driven by users of websites and apps who are typically asked to give up their geolocation, unique phone identifier and to enter personal details into the app. From this point data can be sold, and as one Screenager in her twenties explained, the price for staying connected with the rest of the world is that you must give up your personal details. She explains that the process is unnerving because "I've no idea who will be using my information."

Unnerving or not, one of the seemingly positive implications associated with Big Data (as part of a late liberal environment) is that targeting individuals through digital fitness apps can potentially reduce health problems at the micro level, whilst simultaneously enabling commercial companies to make money. This might seem like a win, win situation, but scholars such as Nick Fox of Sheffield University, point out that focusing on our own individual health (which has become a feature of our evolving health habitus) means that we do little to identify the broader social, cultural and political dimensions of ill health and the reasons why people may find it difficult to keep up or respond to health messages.

Again, this subtle but crucial point was acknowledged by a small proportion of Screenagers, some of which were concerned that "digital health technologies preach to the converted" writes a female in her late twenties. "Educated people that already look after themselves find health apps very helpful but", she argued "people that arguably could be helped the most (the uneducated) won't be interested." Another Screenager in her early forties agrees: "there's a digital divide where apps and devices are marketed to middle class professionals and this becomes a health issue that no commercial company is vaguely interested in addressing."

Choose to Be Healthy

Whilst the perceptions of our Screenagers are not always harmonious, it is possible to summarise some of the themes that accompany discussions of digital health from a socio-cultural perspective. First, participants describe a habitus where agents have become preoccupied with their own feelings, interests or situations—which in turn, reflects the way that people address their own bodies and by implication, health. This movement is thought to be an extension of the Culture of Narcissism that Christopher Lasch described in 1980, but its advancement has intensified with the onset of the internet, and Web 2.0 more specifically.

Interactive websites, smartphones, selfies, social networking sites, and personal apps are transforming the field of health and fitness for everyday people. Most notably participants accept that people have a greater awareness of health related issues than previous generations, and this was largely perceived as a positive outcome of the digital age. For participants, free access to information had begun to change the way that they manage personal health issues, and this was reflected in their thoughts about the relationship between patients and health professionals.

Participants shared their experiences which ranged from: the use of online information to instigate educated discussions with GPs to using personal research in order to correct misdiagnosis from professionals.

For our Screenagers then, the internet has become a tool against which all professional opinion and diagnosis can and will be checked. As such the rise of the net has meant that we tend not to rely on GPs and medics, but we self-diagnose and self-medicate more. Again, this is consistent with self-absorption and is reflected in our attitudes towards exercise, diet and lifestyle.

As well as being cited as a positive experience for the layperson, our Screenagers have largely accepted that the availability and desire for health related information brings with it a heightened level of personal responsibility for our own health and for that of our dependents. In other words, participants were aware that we have become a society of self-monitoring subjects, who are expected (perhaps unreasonably) to self-regulate in a market that paradoxically encourages unfiltered consumption and self-indulgence.

Achievement of self-regulation (which, in the mind of Screenagers tends to equate to staying slim, young, beautiful or athletic) is rewarded with positivity (though sometimes envy) from a society that values aesthetics. As the logic goes, because of the omnipresence of digital technology and e-information, fitness is perceived as a choice where people can, as one Screenager puts it "choose to be healthy or choose to be a burden."

This state of mind, though not a universal position, features predominantly enough to make the list of implications and it's a notion that other scholars have noticed too. For instance, Alan Petersen observes in his 2003 work on the subject of governmentality and medical humanities, that the expectation on contemporary citizens is that they must, as a condition of access to healthcare services, adopt appropriate practices of individual healthcare prevention. Moreover, participants have acknowledged that a heightened sense of responsibility (embedded within the health habitus) can lead some agents to develop negative attitudes towards other people who don't exercise, or are perceived to be lazy, or overweight.

Consuming the latest health-related gadgets and gismos was one option that some Screenagers took to avoid feeling the disapproving glare of the panoptic eye. They understood that fitness apps have the potential to be personal trainers to all: telling us what we ought to aspire to, advising us what think, plotting how much exercise we've done against what it thinks we ought to have done, informing us of what to eat and what to avoid eating, measuring our physiological response to exercise, and of course—photographing our bodies so that we can showcase the end results to our friends, colleagues and to anyone else that cares to pay attention and agree about our social worth.

References

Andrejevic, M. (2005). The work of watching one another: Lateral surveillance, risk and governance. *Surveillance & Society, 2*(4), 479–497.

Ayo, N. (2012). Understanding health promotion in a neoliberal climate and the making of health conscious citizens. *Critical Public Health, 22*(1), 99–105.

Balram, N., Tosic, I., & Binnamangalam, H. (2016). Digital health in the age of the infinite network. *SIP, 5*, 1–13.

Bary, A. M., & Yuill, C. (2017). *Understanding the Sociology of Health.* Los Angeles: Sage.

Bourdieu, P. (1977). *Outline of a Theory of Practice.* Cambridge: Cambridge University Press.

Bourdieu, P. (1984). *Distinction: A Critique of the Judgement of Taste.* London: Routledge.

Buckingham, L. (1998, July 4). Profit from the body beautiful. *Guardian*, p. 24.

Carter, J. (1979). The National Malaise Speech: USA President, Jimmy Carter's Warning of Consumerism. *YouTube.* https://www.youtube.com/watch?v=UZ_hIdOyIas.

Collier, R. (2014). Rapid growth forecast for digital health sector. *CMAJ, 186*(4), E143–E144.

Correspondent. (1979, August 14). Has the ego trip gone too far? *Guardian*, p. 7.

Crawford, R. (1980). Healthism and the medicalization of everyday life. *International Journal of Health Services, 10*(3), 365–388.

Eco, U. (1986). *Travels in Hyperreality.* San Diego, CA: Harcourt Brace Jovanovich.

Erlichman, J. (1992, August 1). The guaranteed truth-free diet. *Guardian*, p. 18.

Fox, N. (2017). Personal health technologies, micropolitics and resistance: A new materialist analysis. *Health, 21*(2), 136–153.

Giddens, A. (1990). *The Consequences of Modernity*. Cambridge: Polity Press.

Guild, W. (1971, May). Fitness forever. *Vogue*, p. 172.

Hardy, M. (2013). In J. Gabe & L. Monaghan (Eds.), *Key Concepts in Medical Sociology* (2nd ed., pp. 133–136). London: Sage.

Jobey, L. (1983, April 3). Another day another diet. *The Observer*, p. 27.

Johnson, D., & Lowe, B. (2015). Emotional support, perceived corporate ownership and scepticism towards out-groups in virtual communities. *Journal of Interactive Marketing, 29*, 1–10.

Knapton, S. (2017, February 21). The 10,000 steps a day myth: How fitness apps can do more harm than good. *Telegraph*. Available at: http://www.telegraph.co.uk/news/2017/02/21/10000-steps-day-myth-fitness-apps-can-do-harm-good/.

Kennedy, J. (1960, December 26). The Soft American. *Sports Illustrated*.

Landesman, C. (1995, March 31). The Bloke. *Guardian*, p. 7.

Lasch, C. (1980). *The Culture of Narcissism: American Life in an Age of Diminishing Expectations*. London: Sphere Books.

Lash, S. (2002). *Critique of Information*. London: Sage.

LeBesco, K. (2011). Neoliberalism, public health, and the moral perils of fatness. *Critical Public Health, 21*(2), 153–164.

Lowe, B., Fraser, I., & Souza-Monteiro, D. (2015). A change for the better? Digital health technologies and changing foot consumption behaviours. *Psychology & Marketing, 32*(5), 585–600.

Lupton, D. (2014). Towards a critical perspective on mobile health and medical apps. *Societies, 4*, 606–622.

Lupton, D. (2016). Towards critical digital health studies: Reflections on two decades of reflections on two decades of research in health and the way forward. *Health, 20*(1), 49–61.

Lupton, D., & Jutel, A. (2015). It's like having a physician in your pocket! A critical analysis of self-diagnosis smartphone apps. *Social Science and Medicine, 133*, 128–135.

Maguire, J. (2008). *Fit for Consumption: Sociology and the Business of Fitness*. London: Routledge.

Mazel, J. (1982, October 5). Slender success or is it just fat chance? *Guardian*, p. 8.

McVicar, J. (1993, October 24). Gym and tonic. *The Observer*, p. C11.

Parker-Jervis, G. (1988, April 10). Fitness and wealth. *The Observer*, p. 67.

Petersen, A. (2003). Governmentality, critical scholarship, and the medical humanities. *Journal of Medical Humanities, 24*(3/4), 187–201.

Prainsack, B. (2014). The powers of participatory medicine. *PLoS Biology, 4,* 1–2.

Probert-Jones, C. (1989, December 3). A healthy diet of yin and yang. *The Observer,* p. 52.

Rees, T. (2017). *The race for the café: A Bourdieusian analysis of racing cyclists in the training setting* (Unpublished Ph.D. thesis). Teesside University, UK.

Robinson, L. (2003). Is digital health technology empowering patients? *Journal of Medical Radiation Sciences, 60,* 79–80.

Stern, M. (2008). The fitness movement and the fitness centre industry 1960–2000. *Business and Economic History On-Line, 6,* 1–26.

Thomas, G., & Bond, D. (2014). Review of innovations in digital health technology to promote weight control. *Current Diabetes Reports, 14,* 1–10.

Trommer, G. (2015). Digital health: Bridging the gap between promise and reality. *Bio-medical Instrumentation & Technology, 49*(3), 182–187.

Utton, D. (2014, June 5). Why the gym is the worst place in the world. *Telegraph.*

Veblen, T. (1925). *The Theory of the Leisure Class. An Economic Study of Institutions.* New York: Mentor.

Yang, H., Carmon, Z., Kahn, B., Malani, A., Schwartz, J., & Volpp, K. (2012). The hot-cold decision triangle: A framework for healthier choices. *Marketing Letters, 23,* 457–472.

Younge, G. (1999, November 15). Fat is a pessimist's issue. *Guardian,* p. B2.

11

Dating

Internet dating is just a modern version of the matrimonial agencies that have been in existence since the 1700s, asserts Professor Harry Cocks [sic] in his 2009 publication, *Classified: The Secret History of the Personal Column*. In this colourful appraisal of the history of romantic relationships Cocks reveals that social networks, created for the purpose of romance, were put to work over 100 years before the birth of the modern internet. For example, he recounts how nineteenth century pioneers such as the forward thinking journalist, W. T. Stead saw the potential of social networks for facilitating marriage between the thousands of young men and women cut adrift in the anonymous modern city. Unlike previous generations of men and women who, in pre-industrial tight knit communities could rely on structured personal introductions to one another through family ties, those seeking partners in the new concrete jungle soon realised that the rules had changed.

Recognising an opportunity and a gap in the market, Stead founded *The Wedding Ring Circle* in 1898 and for 12s and 6d agents would have access to a photo album containing member's pictures and could use the network to look for a husband, wife or friend. Cocks describes how each member would be required to fill in a form that would reveal

© The Author(s) 2018
E. Cashmore et al., *Screen Society*, https://doi.org/10.1007/978-3-319-68164-1_11

details of their personality and their personal requirements. You've heard this somewhere before, right? Thereafter, they received the WRC's monthly magazine, *Round-About* where members could contribute essays on all kinds of subjects, and there was also the opportunity to advertise one's desire for companionship. Take, for instance the following advertisement published on the 2nd July 1898.

> Age 25; fair, slight, fond of music, and a lively temperament; would like to make acquaintance of an educated, refined man not under 30; not necessarily for marriage; wishes to correspond with a gentleman who is cultured and of a sympathetic disposition, either a business or professional man, but must be at least 30, and not more than 50; not a clergyman; a man of broad views and fond of music.

Before making comment on the personal advertisement above, let's pause for one moment to bring the dating story up to date. Consider the following Match.com profile of a woman in her mid-thirties from the year 2016. Does what you read sound familiar?

> **I'm open to trying new things! Email me and I'll respond.** I like to read, and write. I'm looking for someone who can share their thoughts with me, and who won't be afraid to tell me if something is wrong. I'd like to find someone who will mean a lot to me. I admire someone who can discuss their opinions with me, but isn't afraid to disagree. I look forward to meeting you!

Despite the 118-year time lag between extracts, it's undeniable that those advertisements outlined above share similarities. Both advertisements feature women at similar stages of life setting out their hopes and aspirations for a meaningful relationship. The mode of engagement differs (classified column versus internet dating site) but the sentiment is remarkably similar given over a century of change in cultural and literary etiquette.

For Professor Cocks and for some Screenagers too, using forms of new media for the purpose of dating is nothing new after all. One participant in his early fifties from Scarborough explained: "the truth is that dating like this has been going on for decades. Anyone heard of lonely hearts columns?" This participant is right of course, but according to Cocks he is understating the history of this mode of practice, with aspirational relationship advertisements spanning centuries, not decades.

Cocks explains it was only a few decades after the inauguration of the first modern newspaper in 1690 that people would use print media as a means to find a partner, though at this stage openly advertising to seek companionship wasn't accepted as good practice. In fact, reaching adulthood without a partner (derived from 'conventional' communications) was considered shameful, and so whilst matrimonial agencies (i.e. businesses that endeavour to introduce men and women for the purpose of marriage, dating or pen pals) were used, they were rarely spoken about in public discourse other than to disparage users.

Professor Cocks recounts how in Britain, the personal column was suspected of harbouring all sorts of scams, perversities and dangerous individuals, much like the internet is now. He explains how the police were mindful that personal ads might be placed by gay men or prostitutes—both of which were frowned upon as agents of ill repute and so advertisements were watched carefully by law enforcers (*Note* homosexuality was illegal in Britain until 1967; Prostitution is not in itself illegal, but a string of laws can criminalise associated activities).

By way of example, Cocks relays details of the prosecution of Mr. Alfred Barrett, founder of *Link*, the first magazine devoted towards advertisements for lonely hearts in 1915. In 1920 Sir Basil Thomson (Assistant Commissioner of the Metropolitan Police) received a complaint from anti-prostitution campaigner R. A. Bennett (also editor of the newspaper *Truth*) who had sent them a marked up copy of *Link*, with the 'most dangerous' advertisements circled, adding the words "for the benefit of the police." Some extracts included an advert from Iolaus, 24, who described himself as "intensely musical" and of a "peculiar temperament." He had been "looking for many years for a tall manly Hercules." Another, 26, was also "seeking a man pal, who was brilliant, courteous, humorous…and masculine" (pp. 4–5).

The police were aware that code words would act as indications of homosexual interest including "musical", "artistic" "peculiar", and so Bennett pointed out that these advertisements were clearly breaking the law on two counts. First, sex between men was illegal at this time and secondly, it was also illegal to attempt to arrange it. After a court case presided over by Mr. Justice Darling (and bringing to the fore other evidence linking the magazine to other lude acts) Barratt was sentenced to two years in prison in 1921 for corrupting public morals.

Notwithstanding the suspicion that personal advertisements have endured across time, there have always been people that are willing to advertise themselves and others that are eager to reply. The reasons for this were and continue to be multifaceted. They include (though not exclusively) making friends, meeting lovers, and forging communities for minority groups (e.g. LGBT communities).

With his work grounded in historical research, let's assume that Cocks is right and online dating is simply the next logical step in the evolution of dating methods. Even so, whilst it's a finding, at an applied level and in the contemporary setting it offers little in the way of practical or theoretical sustenance. After all, it's the speed of technological change and its applications within the 'real world' settings that requires more thought as the digital turn demands new, more critical perspectives and substantial innovative thinking.

According to Steve Redhead in his 2017 book *Theoretical Times*, as we come to terms with a life lived online, the frantic search for theory is beginning all over again as scholars and social commentators begin to tear apart what they had once settled on as satisfactory explanations for all sorts of phenomena, and this includes the study of romantic relationships. Indeed, for Jo Barraket and colleagues from the University of Melbourne, Australia when writing in 2008, there is an absence of critical academic research on the subject of online dating, and this is despite its prevalence in late modern society. Unanswered questions include: How has the internet and associated mobile devices influenced the way that we think about dating and what consequences does this have for dating practice in the moment and into the future? How far does our approach towards dating reflect other contemporary trends? This chapter aims to find out more.

Fishing

Dating under its various alias's is as old as time itself. Generally speaking (though not exclusively) it refers to a stage of romantic/sexual relationship building where humans engage in communications with one another, perhaps meeting socially (in one or multiple platforms) with the aim of each assessing the others suitability as a prospective partner in a more committed relationship or marriage. (We'll get to *courting* shortly.)

One of the consequences of the advancement of information and communication technologies is the use of the internet, email and mobile phones for the purpose of dating, and according to Screenagers, the concept of online dating is a "game changer because it extends our 'traditional' networks". This Screenager continues: "you don't have to settle for people that you've known through school or work. In a way it's (dating) more exciting and less competitive with your girlfriends who would otherwise be fishing from the same pool."

Here, the participant in her late twenties, raises two main issues regarding internet dating: (1) It can extend social networks to allow for the establishment of new patterns of activity; and (2) It can relieve the pressure or competitive nature that some people feel when looking for a partner from a relatively closed group of known associates. On the latter point, one female participant from Boston, US, reaffirmed: "it (internet dating) takes away the 'left on the shelf feeling' and girls can experiment in order to find what it is they are looking for without feeling like a failure and hooking up with the only available guy in your area." Thus, a common theme affiliated to participant perceptions of internet dating involved the potential to extend social networks.

Writing at the end of the twentieth century, sociologist Manuel Castells observed that "dominant processes are increasingly organised around networks" (p. 469). You'd be forgiven for thinking that this was another case of a scholar stating the glaringly obvious, but in fact Castells was onto something more interesting. While it's true that networks (simply put, a collection of human agents that are connected to one another in some fashion) have existed across time and space and have shaped dominant processes along the way, there was something different about the networks that Castells was referring to. He explained that largely by design, the technological revolution had compressed time and space in relation to communications making it possible for a different type of network to flourish. Where innovation of transport in the nineteenth and twentieth centuries had made it conceivable that people could travel and in doing so, increase social networks (albeit with substantial cost attached), the internet had made network connections possible for the majority of citizens with relative ease, without the need to travel at all, and with only minimal cost attached.

According to Castells, global networks have emerged with profound implications. With an analytical eye on macro level (large-scale) structures, he explains that it is now *information* rather than energy that drives economic activity, with the flow of capital, commodities, stocks and shares dependent on fast moving information about social trends, weather, politics etc. And yet, beyond the movements of the global economy and other forms of macro politics/business, it is important to note that internet networks have had profound implications for individuals at the most basic but fundamental aspects of human existence—including the search for love/companionship via the mechanism of online dating.

It appears then, that networks have penetrated public consciousness and as John Law observed in 2006 when writing about the social study of technology, we seem to have reached a point where every man, woman, child and dog are talking about networks as a means to explain social communications. In relation to online dating, the logic is simple: the internet bolsters our chances of finding a good relationship by usurping geographical constraints and placing in contact with one another, individuals that would otherwise never meet. One male teenager from Philadelphia articulates this very point:

> We are no longer constrained by the availability of potential partners in our immediate geographic area or social circles which are defined by family, work and geography. We create our own circles around common interests … The internet increases our pool of potential mates.

It is worth noting at this point that participants were not suggesting that they would seek out relationships with people from different countries, regions or even towns or cities, but rather that internet dating could free people from the constraints of the tight knit social group that dictated romantic possibilities for all agents prior to the internet age. As one guy in his early sixties from Liverpool explains, prior to the internet age people would attempt the same practice of extending social networks "by attending dance halls, pubs and clubs in the hope of meeting someone new … the internet has just made things much easier."

In the 2005 book *Double Click*, sociologist Andrea Baker was exposed to similar viewpoints as she gathered the thoughts and experiences from 89 couples that met online and started a relationship. One conclusion drawn by Baker was that physical proximity is still a consideration for people when using online dating technology. In agreement, one teenager from London outlined: "the internet is useful for screening purposes, but everyone wants a physical relationship in the end … Trust me, travelling for hours to meet is no fun and won't last."

Similar findings were revealed by Jo Barraklet and colleagues, who in a 2008 study involving interviews with 23 online daters, argued that the importance of exceeding the bounds of proximity is not simply a matter of distance, but a matter of access to diverse networks (p. 157). The majority of dating sites and apps facilitate this function, providing the perfect conditions for users to widen the net in their search for the perfect relationship.

Emotional Scientists

According to Mark Vernon, author of the 2013 book *Love: All That Matters*, our idealised notion of romantic love is actually the biggest enemy of long lasting relationships. He explains that it is socially corrosive because it idealises love and encourages agents to seek perfection in their partners. In other words, we let dreams of romance guide our practical actions. But according to Screenagers, online dating has contributed to a situation in which notions of romance (often attributed to a strong *emotional* desire to connect with another person) are somewhat nonsensically intertwined with science (an *impersonal* and *systematic* enterprise that builds and organises knowledge). For example, a woman in her early fifties from Sheffield explains: "The internet is a more scientific way of looking for the right person … You used to have to kiss a few frogs before meeting your prince, but now, you can rule them out before any kissing takes place."

Others too discuss the merits of scientific dating. A female participant in her early twenties from Leicester observes: "through the internet you can check on how well someone matches with your values, hobbies etc. before meeting them." She insists that this will ensure that "a relationship may be based on substance rather than superficial lust." A male participant in his mid-thirties from Amsterdam, Holland agrees: "on dating platforms the user can input their interests and hobbies to find a better match." He reminds us: "'traditional means' such as meeting someone in a bar is completely based on aesthetics." This point was reiterated by a male participant in his early fifties from London:

> Having fun and meeting in bars is risky. It's not the best basis for a long-term relationship, whereas taking the time to reflect on your own long term wishes through a questionnaire, and then looking through and comparing profiles of potential matches helps to clarify what is really important.

Common to all responses above is the attempted minimisation of risk, a symptom that according to Ulrich Beck in his 1992 book, *Risk Society*, permeates all human beings as the result of implications associated with the new electronic global economy. He explains that risk, in the late modern era, differs significantly from hazards that humans have encountered throughout history. For example, in a high scientific, technological, commercial and largely secular society people accept that humans are masters of their own destiny. They understand that fate is manufactured rather than written in the stars, and so 'risk society' is a by-product of the hyper awareness of decision making that consumes all agents, all of the time.

Beck works on the premise that whilst ignorance can be bliss, circumstances of late modern life (particularly access to the internet) dictate that ignorance does not readily feature. Alternatively, whilst knowledge can be empowering, it also evokes responsibility and an awareness of risk. Consequently, as 'traditional' ways of doing things give way to 'progressive' alternatives people must navigate through choices and manage risks as or before they emerge.

Love by Numbers

In the world of online dating, algorithms are king when it comes to risk management. Algorithms are a set of rules to be followed in calculation or other problem solving operations, and are especially associated with computer use. For example, when setting up an online dating profile, the user inevitably provides information about themselves, and this information is then used to set up the instructions that the computer follows when making a match prediction. Date of birth, favourite colour, pets, medical history, interests, religious beliefs and many other nuggets of information are used by the computer to increase chances of finding a better match and thus minimising the risk of a disaster date. One participant in her late thirties from Illinois believes that algorithms work:

> Internet match-making helps avoid many of the problems which cause couples to break up or divorce. Money, religion, family desires, career goals are all filtered for at the beginning. This saves a lot of effort that used to be spent getting to know people, and then only afterwards finding out they have a basic but fundamental incompatibility with your life plans. I think it is more efficient and it takes the stress out of dating.

This common view is held by many Screenagers who acknowledge two main points. First, that the process of screening potential suiters has revolutionised the concept of dating. Secondly, it has encouraged users to be more open about their feelings, life goals and relationship aspirations in the accelerated search to find Mr. and Mrs. Right. A female participant from Plymouth explains: "people are more open to disclosing information and connecting with people behind a screen." Another female adds: "it's easier and quicker to get a summary of someone online", while one guy explains: "I met my husband on gay.com we put our cards on the table straight away and it worked for us. We've been married 10 years now."

The general philosophy followed by participants like these resonates with the work of scholar Anthony Giddens when building his theoretical argument of the changing nature of intimacy in 1992. As Giddens describes it, modern love is 'confluent love', which means it's dependent on lovers opening themselves up to one another entirely and completely. In essence, Giddens' work is based on a continuation of the arguments about risk and trust that are central to his book *The Consequences of Modernity*, 1990. As Giddens sees it there are contradictions in late modern sexual relations that are implicitly linked to the manifestations of reflexive modernisation because they expose the uncertainties of modern living. His logic is as follows: The sexual revolution (enabled by contraceptive technology and the feminist movement) has levelled the playing field somewhat. Claims to "power and manliness", he writes, "depends upon a dangling piece of flesh that has now lost its distinctive connection reproduction" (p. 153). In other words, the dating game has changed as both men, women and non-binary agents now search for what Giddens terms the 'pure relationship'—based on sexual and emotional openness and equality.

Whilst Giddens is criticised by some for overstating the grounds for equality to exist, it should be noted that his sociology is alive to the fact that not all actors are equal. For instance, in his 1984 seminal work *The Constitution of Society*, he explains how social positioning is always relative to authority within a cultural ream. This might relate to the amount of money an agent has, desirability of their house or number of cars; but equally it can also relate to genetic composition, where agents can be positioned according to race, disability, or in this instance, gender.

Notwithstanding this, it's the seemingly trivial or 'inconsequential' change in thought processes that concern him the most. Seeking perfection from our significant others is a demand that is placed on all agents of late modernity, he asserts. In seeking perfection, we crave continued and consistent stimulation. As a consequence, people view relationships in terms of emotional gratification and this has implications for perceptions of relationship longevity. When speaking of the pure relationship Giddens writes, it is only continued "in so far as it is thought by both parties to deliver enough satisfaction for each individual to stay in it" (p. 58).

With this, it's worth reminding the reader that conceptual love and emotional relationships do not operate independently of social changes. They are in fact part of those changes. Individualization, coupled with contraceptive technology and other social metamorphosis described above have broken down the security once provided by the romantic definition of love 'till death do us part'. This notion is outdated because of the radical overhaul of the societal values that gave it its vigour and importance in the first instance. In other words, love and emotional relationships tend always to track social change, and this was a viewpoint that clearly resonated with a number of cynical Screenagers who discussed the merits and pitfalls of love by numbers.

One man from Darlington commented: "internet couples are likely to have met quickly and therefore the logic is they will separate quickly." Another male added: "with separation and divorce rates never higher, it seems hard to believe that our parents' generation were poorer matches than couples today", whilst one Londoner drew on personal experience to explain: "I met my wife before the media boomed and we've been married nearly 30 years. We managed expectations on both sides, but now it's a case of "Don't like that" and he or she is discarded. People throw things away too easily. It's what life has become."

Fast Love

"I think online dating takes away the friendship building part of the relationship as you are immediately targeting love or acting on lust when you first meet" writes a middle-aged man from Middlesbrough. "Fast love" is how a similarly aged Australian woman refers to this process which she defines as "speeding up relationships so you hit early relationship highs but also early relationship endings." The phrase is from George Michael's 1996 track *Fastlove*.

Those views outlined above, speak of a society that sociologist Zygmunt Bauman would recognise as 'liquid modern' in his attempt to explain an all-pervasive way of life that incorporates and stretches beyond the symptoms of late modernity as outlined by Beck and Giddens. In his 2007 book *Consuming Life*, Bauman contrasts what he considers as the liquid conditions of the contemporary world with a bygone era of solid modernity. To explain, Bauman states that, liquid modernity reflects a society "in which social forms (structures that limit individual choices, institutions that guard repetitions of routines, patterns of acceptable behaviour) can no longer (and are not expected to) keep their shape for long, because they decompose and melt faster than the time it takes to cast them, and once they are cast, for them to set" (p. 1).

Here, he is explaining that the heyday of solid modernity (where society's cultural traditions were rigid and agents knew their place) has passed. Those strong roots and heavy foundations that have characterised the solid period are swiftly diluting and changing shape as they increasingly become liquefied. So, like other theorists such as Anthony Giddens and Ulrich Beck, Bauman explains that liquid life is a precarious life, lived under the conditions of constant uncertainty and concomitant risk.

Moreover, he explains that liquid life is characterized by consumerism, based on fears of being caught napping, of failing to catch up with fast moving events, of being left behind, of overlooking use-by dates, of being saddled with possessions that are no longer desirable. In essence he argues that social attitudes and practices have altered to reveal a recurring succession of new beginnings and swift and painless endings, a situation where getting rid of things takes precedence over their acquisition.

Indirectly, Screenagers have told us that online dating is a derivative of liquid life, given that networks "make life easy for hook-ups and finding friends with benefits", or "speeding up the relationship process" and for keeping romantic options open should something go wrong, or if a relationship has reached its sell-by date. One guy in his sixties from London explained: "nowadays people break up and give up on relationships because they've always got the internet to fall back on. They can be devastated one minute and back in a relationship the next."

All of this was clear to singer/songwriter George Michael (1963–2016) back in 1996 when he wrote the propositional lyrics "Baby I ain't Mr Right / But if you're looking for fastlove…" In his artistic social commentary, Michael drew attention to the changing attitudes towards love and relationships even before the internet created the conditions that would intensify liquid love (from Michael's *Fastlove*).

Screen Dumping

Accordingly, online dating networks can reduce the anxiety or fear of being alone, but also (through consumer choice) can tempt individuals to seek new exciting relationships—which may result in shallow but passionate affairs that rapidly age and expire before each partner moves onto the next. Another relationship is always in reach and according to C. J. Pascoe, when delivering her keynote speech at a 2009 conference about sexual literacy in a digital age, "internet dating is the safe option". She means that the internet can act as a face-saving mechanism that masks the traditional pressures that people (largely males) feel when building up the courage to approach a woman (as Pascoe puts it, though this would obviously depend on one's sexual preferences) with the proposition of a romantic date.

She writes: "men feel less exposed because they can text a girl or leave a message for her on a MySpace page rather than risk the embarrassment of calling her and stumbling over their words or saying something embarrassing" (p. 143). Proceeding in this way may feel less vulnerable, she asserts, because the consequences of rejection are minimised. Equally, as Ilana Gershon explains in the 2010 book *The Breakup 2.0: Disconnecting Over New Media*, young people often negotiate break-ups via a series of conversations across multiple media, a process that might be aptly named 'screen dumping'.

Notwithstanding the utility of the internet as a catalyst for relationship genesis or termination, it was also cited as a potential cause of relationship break down in process. "The internet is your personal tracking device", said one guy from Edinburgh, whilst another from Liverpool added: "the internet can be constraining for relationships. You can see each other's every move and that's not healthy. Everybody needs some down time."

Requiring down time, but being unable to find it is now a feature of contemporary life according to professor Steve Redhead of Finders University, South Australia. In fact, he explains how a life lived online can lead to a condition that is unique to the internet age, 'claustro-politanism'. Entertaining the idea that because privacy is restricted for Screenagers, individuals feel encased, trapped, cocooned or tangled within social networks. There appears to be no escaping the process of lateral surveillance which stimulates feelings of claustrophobia as agents negotiate the all-consuming screen.

All Consuming Love

"Online dating is OK if you can afford the fees", says one Screenager in his late forties. Whilst his point is short, he raises an important issue. Internet dating sites are not free. In fact, of the biggest UK websites that were active in 2017, Emma Munbodh, correspondent for *The Mirror* newspaper, writes:

> *Match.com* has 3 million UK members, each paying £29.99 per month (PM). *Eharmony* has 3.5 million UK customers paying £12.95PM for the service. *Match Affinity* reaches 3 million UK members at £44.95PM, whilst *Elite Singles* has fewer users, though they do pay £89.95PM for the privilege. (3rd May 2017, http://www.mirror.co.uk/money/top-10-online-dating-websites-5220768)

Other reputable dating sites are available too, but the point is this: dating companies work like any other commercial business. They capitalise on the anxiety of consumers by providing access to the materials that could facilitate the achievement of dreams, desires, 'needs' and solutions to personal problems. All available at a price.

Whatever the product—flat screen televisions, mobile phones, clothes, holidays and even relationships, the system works in the same way by stimulating the emotive desire to consume, discard and then consume again. It's an impossible cycle to escape because, when stripped back, everything we do is consumption in one form or another (see Chapter 12). In fact, it is such a fundamental part of late-modern existence that *homo-sapiens* have become *homo-consumers*, says Zygmunt Bauman in his 2005 book, *Liquid Life*.

For Bauman, on closer inspection, liquid life reveals a society of consumers that manage to render dissatisfaction permanent and where agents seek in shops (and only in shops) solutions to all problems. With analogies to shopping it would be easy to think that Bauman is talking about something that's relatively trivial. But it's important to emphasise that the term consumer society signifies something much bigger and much more significant than a large bunch of people who like to satisfy desires by consuming. Bauman writes:

> It truly is a syndrome, a batch of variegated yet closely interconnected attitudes and strategies, cognitive dispositions, value judgements and pre-judgements, explicit and tacit assumptions of the ways of the world and the ways of treading them, visions of happiness and the ways to pursue them, value preferences and 'topical relevancies'. (p. 83)

In other words, consumers are what we have become in every facet of our lives—what we believe, what we desire, and what we do. Consumerism cuts through to our identity as individuals and collectives. The things we think and the way we arrange our lives (and impact on the lives of others), we do it all as consumers. Liquid life is one big shopping trip in which agents (as eternal hapless shoppers) are pushing a bottomless trolley. Happy with our lives in one moment and then trying to fix them by consuming in the next. So, for Bauman, consumer society is more akin to societal attitudes towards life than it is about material purchasing. The latter is often the end result of a recurring process of self-scrutiny, self-critique, self-censure, dissatisfaction, and the endless search for temporary solutions to fast moving situations.

If Bauman is right, then we treat love as a commodity like any other to be bought and sold on the open market, and this practice has become commonplace for singles and even for those looking for a way out of current relationships. In his 2003 book *Liquid Love*, Bauman specifically identifies computer dating as symptomatic of what he calls liquid love, arguing that it has transformed romance and courtship into a type of entertainment where users can date "secure in the knowledge that they can always return to the market place for another bout of shopping" (p. 65).

Bauman's theorising is now particularly fitting as customers choose to ditch mainstream dating websites in favour of the newer, slicker and quicker to use, dating apps. "If you don't use dating apps these days, you are either happily married or living in a hedge off the A34", writes Josh Barrie, correspondent for the *Telegraph* newspaper in 2015. According to Barrie, the supply of dating apps, is intrinsically connected to consumer demand, which in turn is connected to the USP of any given product and this reflects the pervasive marketing approach that is explained by Alan Bryman in his seminal 2004 book *Disneyization*.

In the book Bryman uses the template of a Disney theme park to demonstrate how a series of marketing procedures are used to ensure customer satisfaction. He terms one of these procedures 'theming', which explains how businesses can create unique experiences with distinct themes in order to attract trade. Examples might include a Mexican restaurant, an Irish bar, a toy shop, or a safari park.

Similarly, dating apps differentiate themselves from one another with specific USP's. For example, there's *Tinder*, a location-based social media mobile phone app that facilitates communication between mutually interested users. The steps for use are simple. View the profile picture and then swipe right if interested and left if not. Because *Tinder* introduces people based on geographical proximity it is thought to be used for casual hook-ups.

Another app, *Hinge* has been coined the sophisticated *Tinder* in so far as it links users with friends of friends (using Facebook as its underlying database). The app, *Fuzzy Banter* only allows users to see bio-descriptions, whilst *Voice Candy*, relays audio biographies to the user. *Grinder* is a dating app specifically for gay men and 3nder opens up the threesome market by connecting singles with couples or the other way around.

The point here is this: not only do dating apps reflect a business opportunity for the provider, but they are also reflective of a liquid modern way of life for the user where agents treat internet dating sites as if they were shopping at a convenience store. One Screenager in his late twenties from Glasgow writes: "Dating apps are set up like a shop window. On a superficial level you browse at a picture and think I'm attracted to that!" Likewise, another Screenager from Canada argues: "modern life is all about instant gratification which has taken over young people's ideas of a relationship." For some, the concept of the relationship has transformed so much so that it now exists beyond all recognition.

Avatars and Simulations

Whilst Screenagers hold different views on merits or potential draw-backs of online dating, they all agreed that "the dating game has changed" in a variety of ways. To recap, Screenagers have explained that global social networking and dating sites have widened the potential market of romantic matches whilst simultaneously drawing on scientific algorithms to screen and then filter the amplified field (sorting the wheat from the chaff, so to speak). This was often welcomed by Screenagers as a method of minimising the risk of wasting time with unsuitable suitors. Moreover, they have also acknowledged that those dramatic changes to the dating process might have implications for the 'traditional' practice of dating as they know it. One participant from Chicago describes "the death of courtship" (that being the period during which a couple develop a romantic relationship by spending time together) as agents "get to know avatars not people."

This participant raises one of the consequences associated with online dating, as he sees it. Online daters are missing out on traditional courtship by taking what one Scottish participant views as a "sanitised approach and outlook to dating" where people only display simulations of themselves, and begin a relationship by talking to simulations of others.

Of course, you might think that this is how dating has always worked. Initially people present their best side by behaving in a manner that they think will endear them to their potential suitor. But there's something different about what Screenagers are proposing. They are referring to the fact that the courtship stage of any relationship can now take place in its entirety, online, without ever having to meet in person.

Accordingly, one participant in her late twenties from Liverpool implies: "they (online relationships) are fake, just a bunch of people pretending to be what they want to be, not who they are." Another in her early thirties from London regards online dating as:

> a sort of fiction, but one that can have some authenticity/integrity ... I would say that the 'in person me' is different to the online version but not necessarily more or less authentic. I am more confident writing than speaking and I like the fact that my online persona can have the same thoughts and interests but without the same barriers or neuroses.

For French sociologist Jean Baudrillard, what those Screenagers like those above are describing is evidence of simulacra in process. Simulacra typically describe images or representations of something or someone. He writes in the book *Simulacra and Simulation*, that the task before the social sciences is to challenge "the meaning that comes from the media and its fascination" (p. 84). Writing in 1981, Baudrillard's world views were often revered or disparaged as being radical, but with time, his foresight was proven accurate if not a little tame. He argued that the essence of late modern or post-modern life (as Baudrillard terms it) is imagery and he asserts that this dilutes reality so much so that it becomes a simulation or simulacrum of itself. In a pre-internet world he explained that the media provide one overarching example of an institutional agglomeration that destroys any perception of reality. His reasoning is this: any information given about an event will inevitably become a degraded form of that event, consequently serving to dilute the social. Furthermore, this has a knock-on effect: diluted versions become pseudo 'reality' in the lives of social agents. In other words, as 'reality' is removed an infinite number of times from the actual social event, we are faced with simulations of simulations and thus, an utter absence of any reality.

For some Screenagers then (and in the same way that Baudrillard's explains) internet dating (though not exclusively) provides the conditions for simulacra to thrive in our *Screen Society*. And whilst participants recognise that all relationships inevitably move to 'real time' physical interaction to sustain themselves, it is entirely plausible that the initial courtship stage can now proceed without the need for physical interaction. Instead we opt to give our avatars space to become acquainted with one another. Only then will we take things to the next level.

Beyond the Personal Column

This chapter has revealed, through the words of Screenagers, that online dating represents more than the subtle evolution of the matrimonial advertising sites of the eighteenth and nineteenth centuries—as described at the beginning of this chapter. Rather, transformations to the dating world are reflective of accelerated social change in attitudes and actions which signify a liquid life existence. In the manner that Zygmunt Bauman describes, participants were advocates of, or at the very least were aware that online dating required a new skill set in which visual avatar presentation is crucial to the genesis of any relationship. Thereafter, self-scrutiny, self-censorship, the search for fast love, a desire to consume affection on social media platforms, along-side constant feelings of dissatisfaction—tend to characterise the search for love or companionship.

Participants have explained that dating in Screen Society is full of contradictions. It is risk averse in the use of scientific algorithms that will screen potential partners even before avatars meet; and yet it's risky in the sense that it makes transparent our deepest wishes and desires, laying them bare for all to see and extending the field of potential partners outside of known acquaintances. For instance, one female participant in her early forties from Brussels spoke about the risk: "people can claim to be anything online. I would much rather trust conventional methods and have the benefit of peer-group filtering."

Not only are participants risk averse, but they value the idea of convenient love. Users can search for a love connection at a time of their own choosing, perhaps at lunch time in the office, before bed, or scheduled as part of their online shopping order. Screenagers have indicated that convenience brings with it complacency in the sense that users can rely on the internet to find another date once one relationship has been exhausted or is deemed expendable. Thus, relationships are perceived as idealistic and yet temporary. They are free and yet claustrophobic. All symptoms highlighted above are significations of a liquid modern existence lived through the medium of the screen.

References

Baker, A. J. (2005). *Double Click: Romance and Commitment Among Online Couples*. Cresskill, NJ: Hampton Press.

Barraket, J., & Millsom, S. H. (2008). Getting it on(line): Sociological perspectives on e-dating. *Journal of Sociology, 44*(2), 149–165.

Barrie, J. (2015, July 16). 10 of the best dating apps for men. *Telegraph*. Available at: http://www.telegraph.co.uk/men/the-filter/11741296/10.

Baudrillard, J. (1981 [published in English 1994]). *Simulacra and Simulation*. Ann Arbor: University of Michigan Press.

Bauman, Z. (2003). *Liquid Love*. Cambridge: Polity Press.

Bauman, Z. (2005). *Liquid Life*. Cambridge: Polity Press.

Bauman, Z. (2007). *Consuming Life*. Cambridge: Polity Press.

Beck, U. (1992). *Risk Society: Towards a New Modernity*. London: Sage.

Bryman, A. (2004). *The Disneyization of Society*. London: Sage.

Cocks, H. (2009). *Classified: The Secret History of the Personal Column*. London: Random House Books.

Gershon, I. (2010). *The Breakup 2.0: Disconnecting Over New Media*. Ithaca, NY: Cornell University Press.

Giddens, A. (1984). *The Constitution of Society*. Cambridge: Polity Press.

Giddens, A. (1990). *The Consequences of Modernity*. Cambridge: Polity Press.

Giddens, A. (1992). *The Transformation of Intimacy: Sexuality, Love & Eroticism in Modern Societies*. Stanford, CA: Stanford University Press.

Law, J. (2006). Networks, relations, cyborgs: On the social study of technology. In S. Read & C. Pinilla (Eds.), *Visualising the Invisible: Towards an Urban Space* (pp. 84–97). Amsterdam: Techne Press.

Match.com. (2016). Great dating profile—Short, simple and compelling. *Practical Happiness: Practical Dating Tips and Relationship Advice*. Available at: http://www.practicalhappiness.com/contact/.

Munbodh, E. (2017). Top 10 online websites and how much they cost a month. *The Mirror*. Available at: http://www.mirror.co.uk/money/top-10-online-dating-websites-5220768.

Pascoe, C. J. (2009). *Encouraging sexual literacy in a digital age: Teens, sexuality and new media*. Keynote Lecture: Virtual Sex Ed: Youth, Race, Sex and New Media. University of Chicago Section of Family Planning and Contraceptive Research. June 4. Available at: https://hivdatf.files.wordpress.com/2011/03/pascoe2009.pdf.

Redhead, S. (2017). *Theoretical Times*. London: Emerald.

Vernon, M. (2013). *Love: All That Matters*. London: Hodder & Stoughton.

12

Consumption

Choice

We live in a modern capitalist society. We have always been and remain pretty much all corporate dupes one way or another in the ways in which we consume things. The internet has just made this more obvious. (Male, early thirties, Carlisle)

Corporations are always looking to expand their target market to grow their business and develop products or services which will meet the needs of their consumers. If anything, the ability to communicate directly with their consumers is beneficial since they are able to express their likes and concerns in a process of two-way communication that was not as direct before the internet. For example, the internet has actually made it harder for companies since it is easy to read online reviews and see competitive pricing and research a product or service before you purchase it. Younger generations expect products that are not just mass produced but geared to their individual wants and needs. Think how far we have come from the production line where all cars came in black or red to Nike where consumers can now create their own running shoes in the colours and materials they prefer. (Female, late forties, Baltimore)

© The Author(s) 2018
E. Cashmore et al., *Screen Society*, https://doi.org/10.1007/978-3-319-68164-1_12

These two viewpoints amongst Screenagers reflect the changing and contrasting relationship taking place online between consumers and corporations. On the one hand some Screenagers believe the internet has intensified the corporate manipulation of consumers through direct and immediate access, whereas on the other there is the belief that the internet has given consumers greater powers than they have ever had in history. To put this in some perspective, according to the website www.internetlivestats, the number of global internet users at the time of writing (October 2017) was 3.725 billion. To highlight the growth of the internet over a short period of time, www.internetlivestats also explain how 40% of the world's population has an internet connection, up from just one percent in 1995.

Since the turn of the twenty-first century, the ways in which we consume the internet has become a new social norm and impacted upon how people spend their leisure time, how they work, how they communicate with each other, how they are entertained, how they think and how they behave. This has allowed new industries and markets to emerge and has completely changed human behaviour and consumption patterns. One consequence of this is how the digital age represents a complete shift from a traditional capitalist economy to an ever expanding digital one now increasingly transmitted and consumed through our screens. As highlighted by Alan Warde in his 2017 book *Consumption: A Sociological Analysis*, common terms of reference to consumption are increasingly varied, including "Consumer society, consumer culture, consumerism, consumer politics, consumer demand, the consumer attitude, and above all 'the consumer'." This, he goes on to explain, is "part of attempts to understand the contemporary social predicament of postindustrial and late-modern societies" (p. 205). (Warde's use of the word "predicament" suggests he sees this a troublesome and unpleasant state of affairs rather than just a condition of modern society.)

Given the speed in which the internet has become a vehicle of instant consumption, consumers are making a variety of choices every second of every day across most parts of the world. Whilst consumption patterns regarding various topics (gaming, health and dating) have already been covered in this book, the focus of this chapter was to understand the wider consumption practices of Screenagers and their thoughts on the corporatization of the internet through the increasing levels of attempted engagement with current and potential consumers of products and services.

Changing Consumption Practices

Warde illustrates how consumption covers a range of different practices from shopping, recreation and leisure, mass entertainment through to popular and individual pleasures. Although written before the emergence of the internet, Zygmunt Bauman's 1988 book *Freedom* has striking relevance because of its reference to consumer society as the epitome of freedom, choice and self-direction. Most everyday practices require some form of consumption, but it is not just about the purchase of commodities or the demand for goods and services. For example, Warde illustrates:

> consumption as a process whereby agents engage in appropriation and appreciation, whether for utilitarian, expressive or contemplative purposes, of goods, services, performances, information or ambience, whether purchased or not, over which the agent has some degree of discretion. (p. 86)

Of course, the opportunities for online consumption are not globally consistent, with some countries like China blocking its residents from accessing twitter, Instagram, Facebook, Google and YouTube to protect its governmental interests. Despite this block in the world's most populated country, for billions of others it has become an essential feature of everyday modern life. Part of the reason for this is the simple convenience of the internet in accommodating the varied motivations that influence online behaviour, whether that is using the internet simply as a form of leisure, browsing, purchasing, information gathering or communicating with others via text, pictures or videos. For many people who engage in practices like this, the screen provides a convenient platform that does not interrupt the rhythm of their everyday practice. For example, the 2017 article examining the gratifications of using Facebook, Twitter, Instagram, or Snapchat to follow brands by Joe Phua, Seunga Jin and Jihoon Kim refers to how the mass consumption of social networking sites: "enable users to create personal profiles, articulate their identities, connect with other users and brands, and view, share, upload and comment on photos, messages, videos and other content posted on their newsfeeds" (p. 412).

One important feature of online consumption is shopping, with 58% of Screenagers using the internet to engage in this practice. By purchasing through a screen at their own leisure, consumers avoid queues and can have products delivered to a location of their choice (such as at home or work). On other occasions, such as purchasing tickets, holidays and other services, it can just be sent through online to an email account for proof of purchase or printing without having to leave the comfort of their home.

For many Screenagers was the widespread recognition that the ease of online shopping was only likely to grow further as consumption of the internet continues to expand into new markets and populations. This, some suggested, was evidence of a shift towards a digital global economy, including this response from a male in his late twenties from Middlesbrough:

> The ability to download applications on to phones, the growth of company websites, and the increasing discounts being offered online makes purchasing simpler and also reduces overheads for distributors so it is logical to assume that shopping will become more and more of an online thing.

Another male in his early thirties from Birmingham shared similar thoughts:

> Buying online is different than cash purchases. Debit cards and online payment in my opinion gives a sense of numbness to a purchase. £200 in cash makes someone think whereas £200 from a debit card is gone in the click of a pay now button. Despite this, it is definitely the way people will spend moving forward. We will continue to move more and more into a

society where people do not carry cash in their wallet or purse anymore. Instead our debit details will be stored with a range of companies that we consume on a regular basis. (Amazon, Apple, supermarkets, clothes shops, Ebay etc.)

With regards to the consumption of Amazon (launched in 1994), according to Business Insider, in February 2017, the company had a market share of 43% of all online retail sales in the US. This huge level of consumption was also reflected in data published by the website statista.com, which stated how the net revenue of Amazon was US$136 billion in 2016, up from US$107 billion in 2015.

Another example of changing consumption practices is through our engagement with news and current affairs. It has been widely reported that print newspaper sales are in decline across the world and this was reflected with 77% of Screenagers illustrating how they use the internet to keep them up to date with this information. Another changing practice of online consumption is our engagement with television and music. For example, 59% of Screenagers watch television, music or YouTube. Focusing on the consumption of YouTube, since its creation in 2005 it has become a free repository for the video sharing of media content, including music, television, gaming, blogs and other video clips. According to statista.com, one billion hours of content are consumed every single day on YouTube and this captured market is one reason why advertising has become a key feature at the start of every video that is played on this platform.

In the case of consuming music more widely, certain musicians have a significant global following with fans of Katy Perry (termed KatyCats), One Direction (Directioners), Justin Bieber (Beliebers) and Taylor Swift (Swifties) considered the more die-hard and devoted. Indeed, it was reported in 2016 that KatyCats spent over £1600 a year on her merchandise, tickets and music. For other fans, the opportunity to consume online (such as following these stars on social media) has made fandom seem more real, but there are hidden costs to this practice, with engagement with other fans implicating us in indirect commercial arrangements. Thus, for many fans, it is clear that the gratifications they acquire by being a fan outweigh the economic costs associated with this practice. As Warde explains: "Consumption rarely occurs purely for its own sake, but contributes to the delivery of a range of varied rewards" (p. 93). The internal and external rewards that people gain by engaging with screens are obviously vast, but it is clear that some offer greater social rewards than others.

In their 2010 article focusing on brand consumption and individual practice, Sharon Schembri, Bill Merrilees and Stine Kristiansen illustrate how consumers use brands to construct and negotiate their self-identity. This, they suggest, is often engaged in with others who share a similar interest and can subsequently influence their action and identity. Part of the online appeal and promotion of self-identity is explained by Erving Goffman in his 1959 book *The Presentation of Self in Everyday Life* through the illustration of how people have a desire to communicate who they are to others. As he suggested: "We are all just actors trying to control and manage our public image, we act based on how others might see us" (p. 22). Although this was written before the emergence of the internet, for some Screenagers it clearly relates to the opportunities for this to take place in their everyday practice.

Corporitization

The digital era has given businesses a new cost-effective platform to reach a mass market of consumers. Profits are driven by sales and in this new competitive fast-paced environment all types of corporations are using the internet to engage, connect and converse with current and potential customers. From a marketing perspective, this often includes two primary forms: push and pull. Pull marketing involves businesses providing information to consumers about a product or service with the intention of generating demand and subsequently creating brand loyalty. Push marketing involves businesses engaging in promotional activities with the intention being for the consumer to purchase a product or service.

For some Screenagers, this is just a contemporary realization of the global reach that mass consumption of screens provides to businesses, as outlined by this male in his early forties from West Yorkshire:

People need to realise that corporations aren't here to do us any favours, they are here to take our money and they will use any means possible to invade our personal space to sell their products. We were sold the dream of aspiration in the 1980s and the selling of and buying into that dream has become turbocharged with globalisation. The internet has only magnified this further.

This female in her late fifties from New York concurred: "We have been 'corporatized' since the industrial revolution. Social media is just the latest platform for "those that be" to control the masses in a greater hands-on way than was available to them in the past." This level of 'control' referred to here was an issue for some Screenagers who felt that the relationship sought by corporations was sometimes overpowering, such as this response by a female in her early forties from Swansea:

I do dislike and mistrust the practice of corporations persuading us that they are our friends. I worry about the subtle and pervasive nature of corporatizations reaching into our lives ever more, as well as the corporate use of personal information for their benefit.

Part of this has been a significant increase in what are referred to as banner advertisements on web pages, often based on recent searches that the computer internet protocol address (a unique string of numbers identifying each computer that allows it to communicate with other computers over a network) has logged with marketers who then target the consumer with a reminder of their recent search as well as similar products or services that they might think about in the future.

In their 2008 book *Advertising and Promotion: An Integrated Marketing Communications Perspective*, George Belch, Michael Belch, Gayle Kerr and Irene Powell are critical of this method of corporate advertising through its constant presence on platforms regularly used by people and some Screenagers felt that there would be consumers who would fall victim to this new form of commercial pressure from corporations. Sharing his thoughts on this was a male in his early thirties from York:

I consciously try to avoid adverts and do feel it is somewhat an invasion of privacy. As it is always in people's faces it encourages people to spend money with the sometimes misplaced perception that material things will improve their life in some way. In reality this might only be a minor temporary improvement.

This male in his mid-fifties from West Yorkshire shared similar thoughts:

The internet represents a natural capitalist progression for exposing people to consumerism and hyper-demand. We are manipulated into thinking we need all these products.

A term regularly used on the internet is "clickbait", which explains content whose primary purpose is to attract the attention of and encourage users to click on a hyperlink to a particular webpage where they are usually offered something for sale. However, this method is also used by publishers who can charge more for advertising if they can demonstrate a significant number of clicks on their site.

Drawing on her 2015 chapter focusing on digital consumption and marketing, Jessica Carbino refers to the effect of digital consumption creating a new form of modern capitalism and this was reflected by a number of Screenagers including this male in his late twenties from Cardiff:

In a world of capitalism, the freedom afforded to anyone online has alerted the multinational corporations to the potential audience, either through directly or through influencers. This has proven effective, more so than conventional marketing. The risk of falling for this is probably on par with the risk of falling for conventional marketing.

Other Screenagers referred to programs that are available to limit the level of corporate advertising they receive. Indeed, in a 2017 article for the *Guardian* newspaper, Jonathan Haynes and Alex Hern refer to how one in four people are estimated to have used an adblocker on their desktop and one in 10 on their mobile phones. One of those who used this technology was a female in her early fifties from Ontario who commented: "With programs like adblocker we're actually less likely to be overwhelmed with corporate propaganda than when we were unable to avoid those advertisements on television. The internet allows you to make this choice (if people are aware of it)", whilst this female in her late thirties from Texas stated:

> We are free to ignore or engage as we wish when we are online. Unless we are willing to pay subscriptions for every internet service available, I don't see how the sites can avoid monetizing their content through advertising dollars. It can become annoying, but again, we have the ability to "vote" with our clicks or simply end the intent of the message by ignoring it. Failing that get adblocker!

For other Screenagers, it was just an effective business strategy in utilising all available platforms to engage with consumers, as expressed by this female in her late forties from Los Angeles:

> It is good and effective marketing, sales strategy and public relations to use social media, blogs, webinars, and the like to reach customers. That's the world we live in today and it is no different than search engine optimisation and how that is used. Customers expect to be communicated with and reached via the latest technological tools and trends.

Likewise, this female in her early thirties from London illustrated:

> The huge growth in the number of people using social media has not gone unnoticed by corporations obviously. People spend probably more time on social networking sites like twitter and Facebook than any other online site so it is only natural that corporations will try and tap into this environment to convince people to follow the service or brand and once they do they are often hooked in.

Although television advertising remained the dominant form of communicating products and services to a mass audience until 2017, online advertising has now overtaken tv to become the world's largest ad medium. Given the huge growth of various platforms (Google and Facebook alone attract one-fifth of global advertising spending), marketers have realized the significant potential audience within particular social network communities but also within brand communities that focus on specific consumption activities like cooking, motor sports and computing. In the 2017 research referred to earlier by Phua, Jin and Kim, it was stated how 96% of businesses are using social networking sites to market and provide exposure to their brand and products. One of the reasons behind this is the speed in which social networking operates and this allows communication to take place with a large and more responsive audience than traditional media ever could do.

Brands are using online platforms like Instagram and twitter to increase popularity through the opportunity for consumers to like, share, retweet or comment on any post with the intention to engage with consumers, raise brand or product awareness and drive greater levels of sales. The ultimate aim behind this is to encourage consumers to develop social and personal identities through their consumption by joining a particular brand community with like-minded others. This is why influencers, such as Selena Gomez and Kim Kardashian command over £400,000 per post on Instagram. (An influencer is someone who can exert a powerful influence on consumers' purchase decisions.)

Once in this community, there is a stronger chance that consumers will develop a stronger attachment to the product or service, exhibit higher brand loyalty and ultimately purchase these products or services. For some Screenagers, such as this male in his early sixties from Edinburgh, it can become a herd mentality:

> People want to be part of the in-crowd and are completely duped by the notion that using a certain product demonstrates you are part of that in-crowd. Social media simply provides more opportunities to ruthlessly exploit the desperate need of a majority of the population to feel that they "belong" in a brand community.

One strategy that companies have used is to create online communities who have a shared interest in a particular brand (such as a clothing company like Nike or a car manufacturer like BMW). Robert Kozinets in his 1999 article examining online marketing in virtual communities refers to how this can be organised by the people within an online community themselves or by a company or brand. One advantage of this on the internet is the removal of temporal or spatial boundaries and how a collaborative connection can be established where information and products are discussed and shared between the consumer and the company.

In a 2011 article examining the personalities of users engaging with social networking sites, Margherita Pagani, Charles Hofacker and Ronald Goldsmith illustrate how consumers are more likely to support or follow those brands that fit with their needs and preferences. This, they suggest, leads to a positive perception and evaluation of a product or service and encourages greater levels of communication regarding the brand or company. It can also lead to a collective sense of togetherness when a product or service does not live up to what the consumer expected. For example, this female in her early forties from Inverness stated:

> There are many competing entities for our time, attention and money. People talk to their friends online and in real life more. Thus, customers work together and connect more easily to demand better services and products. If my toaster is broken I might moan about it by myself at home, but if go online and find out that everyone's toaster is broken then I can work with them to make a strong point. I don't need to be rich enough to hire a lawyer or literate enough to write a letter. I can work with other people to make a good point and share our skills via the platforms that are available online.

Likewise, this male in his early thirties from London outlined: "If anything, there is greater scope to challenge corporations and hold them to account for their actions. The internet has given greater voice to the consumer, not the other way round."

One aspect frequently referred to giving the consumer a more powerful voice was the consumer-company relationship on social media. Some Screenagers felt that the use of hashtags and general speed in which messages or posts could be communicated online meant that the consumer held the power in the process of communication. By way of illustration was this response from a male in his early forties from Gothenburg:

Social media is fantastic for the consumer, it used to be you had a complaint, you emailed a company and then they never bothered to reply because there was no public forum to shame them. Now when I have an issue with a company I email them, tweet them, Facebook them and 99 percent of the time the fastest response is on social media as companies don't like to be shamed publicly.

Electronic Footprint

Certain companies such as Amazon, Facebook, Instagram, Apple, Google, twitter and YouTube have significant online power that culminates in mass consumption of these platforms. As suggested above, digital information about consumers has become a crucial source of commercial value to companies, particularly when you consider that 70% of Screenagers engage with social media and 65% use the internet for personal research including browsing online sites for news, information, products and services. The increasing consumption of the internet has resulted in a significant amount of growth in the analysis of our everyday practice and the sharing of information around our online interests, tastes and habits. In a 2011 article explaining the emergence of brand communities, Stephen O'Sullivan, Brendan Richardson and Alan Collins outline how our online activity allows companies to track purchasing history, browsing, brand engagement and loyalty. Through these electronic footprints that a number of people unknowingly provide to marketers and businesses regarding online behaviour, tastes and habits (provided via computer coding through what are termed cookies and other forms of spyware that monitor and record most of our online social encounters such as browsing and purchasing choices), businesses can promote new products and services through a targeted approach on regularly utilized online platforms.

As explained by Carbino, "social relationships are a major part of the consumption and marketing process" and the digital era gives businesses a chance to target their demographic interest base by accessing their shopping patterns, transactions and payments and feedback via reviews and chat rooms about products and services (p. 245).

For some Screenagers, this covert surveillance by corporations encroaches too far on the corporate-consumer relationship. By way of illustration was this response by a male in his early fifties from London:

> Stealth marketing through profiling is in my opinion wrong, but it has become more and more prevalent the last few years and I foresee this only expanding. They know where you live, what you use your device for, when you use it, where you use it, how often you use it, who you bank with, what you buy, where you eat, where you shop, where your leisure is spent. A marketer's dream, but not necessarily a consumer's.

Likewise, this female in her early forties from Liverpool stated:

> It is like a covert version of Big Brother that has you under surveillance when you are at your most vulnerable online, such as searching from goods and services or making some form of transaction. It is like stalking but rather than be a criminal offence it seems to be acceptable in the corporate world.

Despite the ethical questions that this provides to corporations and the diverse forms of online communication that exist, as suggested earlier, they are devising many ways to create current and future brand loyal consumers by seeking to understand their wants and needs. It is a tactic that some Screenagers recognize, albeit with mixed opinions as expressed by this female in her late forties from Nottingham:

> I am reluctant to share personal information online, although clearly every search undertaken leaves a footprint from which your interests and lifestyle can be monitored and potentially exploited through aggressive marketing. It is good for business but it really does get annoying, especially when it pops up on sites like Facebook, I feel like that is an invasion of privacy.

This male in his late thirties from Portsmouth added:

> I am amazed how so many people know so little about what Google, Facebook etc. are doing. How much of their lives are being recorded, and then processed for profit-making opportunities. You have to opt-out of

every Facebook encroachment into privacy and it usually takes a scandal before anyone of them retreat slightly. Some of the patents under development by these companies are frankly scary. And what is right about looking for a holiday on a website and then logging on another day to find adverts on the side of the screen when looking for something else?

The relative cost-effective nature of targeted advertising was also raised by some participants, particularly when the product had already been researched online and subsequently purchased. For example, this male in his late thirties from Toronto explained:

> Targeted advertising pisses people off and often happens after a product has already been purchased. Search on Amazon for a barbeque. Buy one. Every advert you see for the next month will be for a barbeque. Er hello, I have bought one - how much money and time are these people wasting on dead leads?

This female in her early fifties from Ottawa shared similar frustrations:

> I am not sure if it's any different than getting inundated with commercials on television, but online it feels a bit different as it is in your private space. It is especially annoying to suddenly have every internet feed clogged with product advertisement. I once searched online on my home computer for an industrial shelving unit for my job. After that it took months to stop seeing shelving ads on my Facebook page, Amazon page and almost all of the search engines I use. Very annoying.

Other participants were more reflective of corporate surveillance, such as this female in her early twenties from Sunderland:

> I think that it is invasive when corporate companies are all over the advertising banners on my social media as a result of my recent search history. However, I suppose that it gives the consumer a chance to interact with and shape the companies to how they want them to conduct their business.

This male in his early forties from Greater Manchester added:

> I have no problem with advertising as it has always existed as long as there are things that companies need to try and sell to a mass audience. It is important for businesses to get as close to their current and potential future customers if they are to survive in this competitive online environment.

It is not surprising given the accounts so far in this chapter that internet advertising has become a significant feature of promoting products or services with the emphasis on the consumer purchasing them. This is not just done through social networking sites and other forms of online advertisements, but also via email communication because of its relatively low cost and ease of access to thousands of people. However, this can lead to unsolicited emails, with the website statista.com estimating that nearly 57% of email traffic is spam. For some Screenagers this has reached worrying proportions, especially as it is becoming increasingly targeted to influence individuals to spend their money in some way. "I am sick to death of email spamming. It is constant, every single day, and it is such an invasion of privacy", said this female in her mid-forties from Leeds, whilst this male in his late twenties from Bristol concurred:

> For the vulnerable amongst us you can see how email advertising would have its advantages. Things like "too good to be true" are usually too good to be true, but not everyone sees through this. It's an invasion of the vulnerable.

Whilst television and online adverts are always consumed visually in the first instance, what online advertising allows is for people to immediately click on any advertising banner or link. Thus, rather than digest the advertisement like we do through television, for some people of the so-called one-click generation, impulse can create a sense of intrigue to find out more about the product or service and this can then lead to an unwanted purchase.

However, other Screenagers like email communication as a form of targeted advertising, such as this male in his late fifties from London:

> I get lots of offers via email communication and I take advantage of them: free meals, cheap holidays, shopping offers, free bottles of wine in restaurants, cinema tickets, theatre, cheap running shoes etc. – and I am thus in

receipt of far more corporate messages than ever. Emails in my inbox have corporate messages but are often graced with offers to entice me to participate and if I think there is something in it for me then I will participate. We always have the option of unsubscribing if these emails become too frequent or not useful anymore.

This sense of control was also reflected by other Screenagers including this female in her early forties from London:

When businesses ask for my email address I ask them how they will use it and how often they will send me marketing stuff. I don't mind a monthly update, but when I feel bombarded I unsubscribe (remember we do have a choice to do this if we feel the need to do so). What does irritate me is how searches have a very corporate, big business bias, due to search engine optimisation. I miss the serendipity of early times on the internet and the ability of smaller businesses to more make more of an impact than they can do at present. It is as if our electronic footprint is for the big corporations to engage with only and everyone else can go and do one as if online territory belongs to them.

References

Bauman, Z. (1988). *Freedom*. Milton Keynes: Open University Press.

Belch, G., Belch, M., Kerr, G., & Powell, I. (2008). *Advertising and Promotion: An Integrated Marketing Communications Perspective*. Sydney: McGraw-Hill.

Business Insider. (2017). Amazon accounts for 43% of US online retail sales. February 3. Available at: http://www.businessinsider.com/amazon-accounts-for-43-of-us-online-retail-sales-2017-2/?r=AU&IR=T.

Carbino, J. (2015). Digital-mobile consumption and marketing. In D. Cook & J. Ryan (Eds.), *The Wiley Blackwell Encyclopedia of Consumption and Consumer Studies* (pp. 244–246). Chichester, UK: Wiley-Blackwell.

Goffman, E. (1959). *The Presentation of Self in Everyday Life*. New York: Double Day Anchor.

Haynes, J., & Hern, A. (2017, June 2). Google to build adblocker into Chrome browser to tackle intrusive ads. *Guardian*. Available at: https://www.theguardian.com/technology/2017/jun/02/google-build-adblocker-intrusive-ads-chrome-browser.

Kozinets, R. (1999). E-tribalized marketing? The strategic implications of virtual communities of consumption. *European Management Journal, 17*(3), 252–264.

O'Sullivan, S. R., Richardson, B., & Collins, A. (2011). How brand communities emerge: The Beamish conversion experience. *Journal of Marketing Management, 27*(9–10), 891–912.

Pagani, M., Hofacker, C. F., & Goldsmith, R. E. (2011). The influence of personality on active and passive use of social networking sites. *Psychology & Marketing, 28*(5), 441–456.

Phua, J., Jin, S., & Kim, K. (2017). Gratifications of using Facebook, Twitter, Instagram, or Snapchat to follow brands: The moderating effect of social comparison, trust, tie strength, and network homophily on brand identification, brand engagement, brand commitment, and membership intention. *Journal of Telematics and Informatics, 34*(1), 412–424.

Schembri, S., Merrilees, B., & Kristiansen, S. (2010). Brand consumption and narrative of the self. *Psychology & Marketing, 27*(6), 623–637.

Warde, A. (2017). *Consumption: A Sociological Analysis*. Basingstoke: Palgrave Macmillan.

13

Privacy

A World Without Screens?

It probably won't happen while this book is still in print, but it *will* happen. It's hard to imagine it now, but, one day, the smartphone will meet the same fate as the Sony Walkman, Polaroid camera, transistor radio and all those other electrical appliances we once marvelled at and assumed we could never live without. We always find a way of living without them.

Today, we look at our smartphones, not so much as revolutionary gadgets, but prerequisites: these are just things that are required as prior conditions for something else to happen; in this case that something else is our ability to get through a normal day. At least, to get through a normal day *functionally*. It's difficult to think ahead and imagine how we will function without portable screens. But, who knows?

Perhaps a quarter-century from now, we will travel underground at 250 miles (402 km) per hour and wonder how we ever struggled with planes and trains. Just like we smile when we think about those fax machines and how we once listened for ear-piercing, high-pitched screeches that indicated the document transfer process was

© The Author(s) 2018
E. Cashmore et al., *Screen Society*, https://doi.org/10.1007/978-3-319-68164-1_13

underway. In the 1980s, when they were introduced, faxes were truly revolutionary. They made airmail irrelevant and many long-distance phone calls unnecessary.

The phones we currently carry around with us, by contrast, are more evolutionary in the sense that they developed from earlier forms of the same technology to change our behaviour patterns without actually making any of them redundant. We still use desktop computers, laptops and regular landline phones; we take photographs on cameras, keep sounds on digital voice recorders and listen to ambient music. And, of course, we still play games on other devices. It's probable that our phones, or something similar to our phones, will soon be controlled solely by voice commands and have some sort of virtual reality augmentation—though those big headsets that project 3D images are already looking a bit clunky. The question is: will screens survive? Well, they have for 170 years or more, though, since the 1950s and the rise of television, they have become progressively indispensable. And indispensable is the right word: screens are absolutely necessary today.

It's been speculated for a while that screens will meet their doom once standalone augmented reality gets into its stride. Headsets and glasses capable of delivering three-dimensional images to the pupils of our eyes have been talked-about and used, though without conspicuous enthusiasm. The concept of dispensing with screens and replacing them with detailed images that are visible and can be acted on is now being discussed. There would be no point in having a handheld gadget or even desktop products for either work or entertainment—or anything else. Calls, games, movies and tv shows could all be visualized and would respond to verbal instructions. No doubt they would also interact verbally. This sounds speculative though not unimaginable. Anything further takes us into the realms of loading data into our neural circuitry and posthuman possibilities that are more properly the subject matter of science fiction.

Voice assistants and virtual realities, including holographic imaging are already available and probable extensions of technology we've already created. But the technology that supersedes screens will probably be something other than a combination of these. For the time being, screens of some sort will populate our world and continue to affect the way we live. As we've tried to show in this book, the dire consequences of depending on screens, particularly on smartphones, have been exaggerated. They're also based on widespread anxiety, if not outright alarm, rather than rational analysis or research; though some research is so fraught with panic that it seems like it is part of the stampede for the exit. The research on which this book is based informs a different type of conclusion: that there is little to worry about.

The only effect of screenlife today we should fear is that we overspend: screens have helped turn us all from consumers into hyper-consumers. We buy stuff every day and chances are we'll keep buying. The difference between today and tomorrow is that we'll probably spend less over retail shop counters and more by hitting the *Pay with PayPal* button.

We're not underestimating the effects of the latest generation of screens on our everyday life; this book plots a course through our activities in order to show how every aspect of modern life has been impacted by screens. We too have changed, though not in the way alarmists assume. This is hardly a profound observation: we are always changing: we're constantly responding to the environments we create.

Technology dominates those environments; it has done since the advent of the steam engine. Innovations over the past 30 years have made an impact as major as anything since the Scottish engineer James Watt's improvement of Thomas Newcomen's pump to remove water from mines in the eighteenth century. Watt's machine didn't change us: we used it to recreate our environment and then reacted creatively to that new environment. That's what is happening today.

A hundred years from now, people will look back and identify minor changes in the way we conducted ourselves, just as we do when trying discern how we changed in the twentieth century. We are changing right now: as we write, we are experiencing changes that will be manifestly obvious in the future, but are only just about apparent at the moment. We will close our account by considering the two changes in us: not just the way we live, our habits, and tendencies, though, of course these are cumulatively who we are. *Screen Society* has changed us as human beings. The two most significant changes, as we see them, are in the ways we relate to each other and in our conceptions of privacy.

Antisocial

"You have to wonder at what point have you grown a generation that cannot interact with anyone else." Newt Gringrich, the US Speaker of the House and former Republican representative from Georgia, was probably echoing the thoughts of countless Americans and people around the world. It was October 2017, ten years after Apple had introduced its first iPhone.

Gingrich was then 74, a couple of years too old to be part of the baby boomer generation, and his views were arguably representative of no one in particular. As the *Screen Society* research indicates, there are no age requirements for Screenagers. All the same, the presumed inability of people—and we assume he meant young people in particular—to talk to people was dividing not only the US, but the world. In particular, Gingrich singled out texting as a major problem: the love affair people have with texting was retarding their communicative skill.

Not that Gingrich's argument is without support from younger people. "We rely on them far too much and they make us antisocial," a man in his early twenties told us, adding the reminder: "But they can heavily aid education so we do also need them now." The tradeoff is worthwhile, he reckons.

Others reckon there is no tradeoff required. People like Gingrich just don't understand that communication itself is changing. It has probably been changing since time immemorial, of course; not just the means or methods of communication, but the content too. Today, we talk to each other differently to how we talked 20 years ago. Watch a 20-year-old film and you will hear unfamiliar words; try to notice the absence of "like" as we use it today (as a substitute for "said" or "thought"); try to see the nonverbal signals and how they differ from the gestures we use nowadays, particularly the facial expressions—people won't draw back their heads and pull that exaggeratedly incredulous expression that silently says, "I don't think so."

Communication is constantly changing, though not in an even way: as with evolution, the modification is constant but irregular. It sounds a contradiction, but imagine that substantial changes wrought by the telephone from, say, 1947, after which there was a rapid rise in the number of homes with domestic phones (the actual piece of technology was available from the start of the twentieth century). Talking to someone who wasn't physically present, nor even in the same city and, after 1956, possibly not in the same country, must have been disconcerting at first and forced an adaptation based on language alone; think about how much we use our bodies, especially facial expressions when we talk face-to-face.

"They [screens] have changed the way we communicate. I believe now screens have caused us to communicate less face to face," said a female participant in her late teens. It seems an obvious point, though think about the second aspect: do we communicate face-to-face less than we used to? Or do we augment our physically proximate interactions with screen conversations? Because the overall impression the *Screen Society* project provides is one of people engaging in both forms of communication, often simultaneously.

A teenage woman enthusiastically welcomed the multiform communication:

> It's a new form of communication, it allows us to sustain relationships that would otherwise perhaps fade due to the effort required to contact people through letters or phone calls. I heard a comment saying that people don't talk or socialize much when they use phones but I think they are still socializing with other people through their phone/laptop or whatever device they are using.

Maybe the term face-to-face, conventionally meaning close together and facing one another, is in need of modification. As another young woman in her early twenties pointed out: "I think we have more of a global society, we can see the faces of loved ones who are miles away."

And, while we are updating concepts, Gingrich's reference to the generation that can't "interact" invites us to wonder whether interaction needs attention. A reciprocal action or mutually influential exchange can be conducted in several different ways. Our immediate thought might be about people standing in each other's presence. But online interaction is still reciprocal and involves polyvalent influences. We use the term polyvalent to mean having different forms, facets or purposes—all implying influence to some degree.

Some Screenagers shared the same reservation as many critics: we presume we have only a finite amount of time available to us and a limited capacity to interact with other humans. For example, one female in her early thirties outlined: "Although we are more interconnected with those far away, our relationships with those close to us is probably hindered by screens."

This is a plausible argument and one predicated on the understandable assumption that there are boundaries that define the limits of human possibility. Even Darwin would have accepted that our capabilities are not infinite. Yet we learn new skills with surprising adeptness. Communication, argued one of our teenage participants, is, itself, learning: "It (the screen) has changed how we learn. News is easily conveyed through social media platforms and apps allowing people a greater diversity in how they understand what's going on in the world." She went on:

> I do think screens have had a negative impact in the sense that people can become too concerned with checking their devices more often than is necessary, and that they have enabled issues such as cyber bullying and cybercrimes to become more prevalent. Yet if people use them safely and properly then I think they have been an amazing contribution to society.

Like every other facet of our being, our capacity to interact with others is in constant evolution. It's easy to imagine the way we greet, mix and socialize with others has remained constant over the decades and even the centuries. We've been persuaded by historical dramas and reconstructions. But we have no way of knowing. The way we address each other, sustain friendships, relate to others, even the language we use in our exchanges are forever changing. People, who, like Gingrich, believe we are losing our ability to interact, fail to see we're doing nothing of the sort: we're just interacting differently and with people we can't always see and in ways that just weren't available to earlier generations. Social interaction on screen involves acquiring the abilities to interpret meanings, often in visually and audibly complex ways. It's interaction, though not as Gingrich et al. know it. It's not antisocial; it's newsocial.

Nothing Is Private

In 1991, NBC introduced *The Jerry Springer Show* to American television: this was a standard fare, daytime talk show for a few years, before it was revamped into a sort of group confessional session, guests were invited to share their dirty little secrets with a studio audience and millions of tv viewers. It became a wildly successful format with the British-born host Jerry Springer inviting guests to reveal and explore unimaginably intimate details. Unimaginably, that is, in the early 1990s. Nowadays, people share without inhibition, suggesting how elastic the concept of privacy has become.

Over 4000 miles away from the Chicago studio where Springer's show was filmed, a Dutch tv producer, John de Mols, must have been watching and learning. People were becoming unselfconscious and utterly relaxed about talking frankly, even about aspects of their lives that made audiences blush. What if you could make them not just talk but behave in a way they would register the same reaction? De Mols, a partner of Netherlands-based tv production company Endemol, seems to have had a Eureka! moment: into his head popped an idea that must have seemed doomed from inception, but which developed so successfully that it changed culture: *Big Brother*.

Privacy is conventionally understood by what it isn't: not observed, not disturbed, not available to or to be disclosed to others. If we want something to remain private, we exclude others, or at least most others apart from a small number of confidantes, from our conversations or other activities. We either keep the private thoughts to ourselves or share them with a select group of others, who will not reveal them. At least, we hope and assume they won't.

What Springer's show and the several versions of reality shows that followed *Big Brother* showed is that privacy is not as valued as it once was. It used to be important to preserve and protect an element of ourselves, to maintain a distinction between what we do and say in public and what we think and feel to ourselves. By the time twitter launched in 2006, viewers of confessional and reality tv shows were habituated to being guiltless eavesdroppers on other people's conversations.

Springer had given impetus to many imitators. Shows purporting to offer help and advice to those with personal problems were effectively acknowledgements that some people were prepared to talk openly about something they were ashamed about—and many more others were willing to listen to them and enjoy the experience. Reality shows were the equivalent of peepholes: viewers were invited to look through the small hole and become a voyeur, if only for the duration of the programme. (There's actually a better analogy and one that takes us back to chapter one, in which we covered the invention of the primitive motion picture machine, the kinetoscope: in the early 1900s, a coin-operated variation called the mutoscope became popular and its most popular narrative was "What the butler saw." The consumer dropped a coin in a slot then looked through a peephole, as if he or she was a butler gawping through a keyhole at the Lady of the house as she was undressing. By 1930s, the entertainment was superseded by filmed pornography.)

At the start of the 1990s, it would have been unseemly to watch an afternoon tv show in which a grown man discussed his premature ejaculation, or a woman who talked openly about her lesbian girlfriend, her smack-addicted husband and her penchant for giving oral sex to strangers in public places. These would have been unambiguously private matters. By the early 2000s, they were the everyday subject matter of several tv shows.

Social media didn't turn us all into voyeurs: it just complemented a process that had already begun. And, without wishing to pause the argument for long, we should add that privacy was in the process of redefinition from the 1960s, when Italy's paparazzi started to take an uncommonly prurient interest in the private lives of the rich and famous—and we, the audience, found we found their stories and pictures oddly rewarding. We agonized, fretted, questioned, praised, damned, approved-of and tut-tutted. And all the time, we felt entitled. Entitled that is to engage in others' lives.

So, in 2006, when twitter delivered a method of engaging in the most direct indirect way, so to speak, it arrived gift-wrapped. Twitter enabled users to share snippets of information about themselves, or others, with potentially anyone else who used the social network. Although the original idea was to allow people to send pieces of information that were as inconsequential as the chirruping of birds, it soon morphed into a gossip medium and, later, a marketing tool for celebrities to peddle commodities.

It's probable no one anticipated how tempting twitter would become. It's not as if people were enticed into sharing confidential information. After all, twitter didn't coax persuade or sweet-talk tweeters into disclosing anything. But it offered a kind of purgatory device: we don't mean it was a place of suffering or torment for those wishing to expiate their sins before going to heaven; but a channel with a cleansing property—people could release whatever information they wanted with impunity (within reason, at least). And, in a kind of self-perpetuating manner, others responded with comparable candour and lack of inhibition to find the release was surprisingly purifying.

All this could tempt us into believing privacy, at least privacy in the traditional sense of the condition, has disappeared. Screenagers insist that it hasn't: it's been changed. "People are relaxed due to social networks," said one man, in his forties, from Scotland; though a woman, in her thirties, from the northeast of England asked: "Is anything really private anymore?" A man in his twenties, also from northeast England had an answer: "Privacy is whatever information you've successfully managed to keep exclusively to yourself. *Nothing is private.*"

Despite this, half of the 303 people who were prepared to write about privacy believed it still existed, though only for those who took care. "It is kind of like passing paper notes in school, and having the teacher intercept a message," wrote an American woman in her fifties. "If we want to limit who can see our data, we need secure ways to share it and both parties need to respect that security." Her point is that potentially we can protect information we wish to remain private. Practically every participant agreed that fewer and fewer of us care very much about protecting what would, a decade ago, be regarded as personal.

The word "relaxed" cropped up time and again, indicating that sharing is not a tense or anxious experience, but one that's enjoyable. The kind of details that would make earlier generations squirm are now the stuff of amusement. "It's no different from walking around one's own house naked all day," said an Englishman in his twenties, implying that this kind of nudity is agreeable. He also added: "with no curtains on the

244 E. Cashmore et al.

windows" to emphasize how one person's pleasure may be the source of another's outrage.

It appears the concept of privacy is still recognized by Screenagers, who believe the traditional binary exists, though with modifications. Many believe social media offers sufficient options for those who wish to limit the amount of information they share. Others doubt anything can remain airtight once it's surrendered to cyberspace: "We all sign our lives away every day." Screenagers live in terror of losing personal information, but it's what we might call a relaxed terror.

Like every other condition, or to be precise, every word that describes a human condition, privacy is constantly being renewed. "Wellbeing" today means something different to what it did thirty years ago. "Depression" didn't exist in the way we use it today prior to the 1980s. "Satisfaction" has always been variable. So the meaning of privacy in the 1990s was different to its meaning in the 1890s.

Our research indicates that Screenagers believe privacy exists; they just think the boundaries are more porous than they were before interactive media made sharing not just easier but agreeable. As we've stated before, privacy is elastic. Sharing information about ourselves is not new; experiencing it as pleasant, rewarding and perhaps even irresistible is quite new. This means Screenagers are comfortable sharing information concerning their private lives, relationships and emotions; they don't consider it inappropriate or offensive to share details of their personal life. They still recognize that this is private, but private no longer means that the information belongs exclusively to one person or group of intimates. It sounds paradoxical; but, not to Screenagers.

The Capture of a Generation

"Our lives are dictated by screens to the extent that our brains' biological response to notifications mirrors that of the pleasure triggered by food or sex." It sounded as if technology journalist Rhiannon Williams was speculating on what life would be like in the not-too-distant future. She was making an observation about the condition evolution has reached in 2017. "For the past 200 years humankind has been hell-bent on creating machines and devices capable of the same tasks and functions the majority of us were born with." To some extent she is right. But there's a McLuhan-ish determinism about her portrayal of technology as a dictator rather than servant of humanity.

Marshall McLuhan was one of the most provocative thinkers of the 1960s. In his influential book *Understanding Media* (first published in 1994) McLuhan argued that our main modes of communication—telegraph, radio, television, movies, telephones and then relatively primitive computers—were reshaping life in the twentieth century. It sounds a reasonable and unremarkable statement today. But, in the 1960s, it was a moral as well as technical pronouncement. Could human-made artifice actually become a dominating force to the point where it changed us? McLuhan's famous phrase "the medium is the message" meant that the characteristics of a particular medium rather than its contents—the information it contains—influence and control society. Put another way: the way we acquire information affects us more than the information itself. Television and radio, the two media in which McLuhan was most interested, were electronic extensions of our central nervous systems. Williams' comment about devices capable of human functions seems to complement this.

McLuhan has been credited with foreseeing the internet, or at least a communications facility consisting of interconnecting systems that resembles what we now call the net. He also used the phrase global village to capture how standardized telecommunication protocols would become accepted worldwide so that we would consider the world as a single community. Yet people are still deliberating about whether his arguments warned of a future dictatorship or promised a utopian society in which we all lived comfortably in a single environment, an environment developed and sustained by electronic media. This was the import of his phrase, the medium is the message—the medium is an overall structure that affects every facet of human activity, including how we think.

One Screenager agreed with McLuhan's conclusions, that we have, in her words, "been taken hostage." Now in her sixties, she has witnessed the capture of a generation by seemingly innocuous devices that now secure them. She wholeheartedly accepted that there is an addiction:

"The need to share the most trivial information with others is inhuman in behavioural terms and has led to many unintended consequences, not to mention crimes." And yet, there is an emancipatory potential: "As an educational tool or information finder using your screen can be the best and most convenient means of doing so."

And it is this point we need to bear in mind when assessing the impact of screens on our lives. Every advance, sideways step or regression, cultural or technological, involves some sort of capture. By applying ourselves we are putting our brains and bodies into operation. In the process, we surrender ourselves. We've done this with cars, planes, and multiple other machines. Works of art don't emerge: they're created, often painstakingly. Nor is enjoying a passive activity: we give our attention, in some cases our full attention, or risk not getting the benefit. In other words, when we think we're living under and held captive by a technological dictatorship, we should remind ourselves that we created it and we use it—not vice versa.

Few people, young or old, do not understand the captivating prowess of screen devices. Our research suggests that younger members are more confident in the ability of humans to resist; they just don't want to. When you think about it, the screen is a piece of brute technology that has, in less than 200 years (longer if you consider the fascination with the magic lantern), found a home in our head and our hearts. As we stated earlier in this chapter, there are signs that it will disappear, but only signs; there is no obvious replacement to those panels that display words and images with such force and effectiveness.

References

McLuhan, M. (1994; originally 1964). *Understanding Media: The Extensions of Man*. Cambridge, MA: MIT Press.

Petersen, B. (2017, October 5). Newt Gingrich says millennials are dividing America by texting too much. *Business Insider Australia*. Available at: http://bit.ly/2knAloy. Accessed October 2017.

Williams, R. (2017, October 11). Remember the joys of humanity? *i*, p. 18. Available at: bit.ly/2ybQF1j. Accessed October 2017.

Methods

The *Screen Society* project is perhaps the most globally ambitious online qualitative academic research project to date. We say this because, as far as we are aware no scholar or research team has attempted to deliver a project on this scale before. In a sense, we've been flouting, or perhaps more accurately, re-writing the rules of qualitative enquiry.

Students on university courses around the world are often told that qualitative research of any value must deal with small sample sizes, and that qualitative surveys are difficult to complete, if only because participants don't have the capacity or desire to respond to research questions that require personal time investment. In other words, their lives are busy and they are less inclined to expound at length. We found the opposite.

People from all over the world took the time to offer their views; in many cases, explaining their experiences, understandings, interpretations and foresight in intricate detail. There was little sign of qualitative fatigue from self-selecting participants.

© The Editor(s) (if applicable) and The Author(s) 2018
E. Cashmore et al., *Screen Society*, https://doi.org/10.1007/978-3-319-68164-1

Of Their Own Choosing

So, how do you investigate how people think and feel about their screens, how they react to them, organize their lives around them, understand their relationships to them? Not with a clipboard and a series of tick-box questions, or a tape recorder and a two-hour conversation (though the latter would probably be interesting). We chose the more obvious method: through screens.

Researchers use the internet all the time nowadays, of course, though they are mostly interested in methods designed for measuring by quantity rather than quality. So their efforts are to produce statistics that can indicate trends or patterns, but without yielding insight as to how people think. All three of us are interested in assessing by quality rather than quantity; so statistics were of no importance. Insights into people's ideas, assumptions, impressions, perceptions, convictions, opinions and, generally, their lines of thought were our priority. People who use their screens don't do so mindlessly: as this book has shown, they think about what they're doing.

Those who agreed to provide their opinions did so via their screens. It didn't matter whether they chose smartphones, computers or tablets. Nor was the environment of any consequence: they chose whether to compose in privacy or in a crowded room; while focusing exclusively on the research, or while watching tv, talking to friends or doing practically anything they wished.

Two of the present trio of researchers devised (or perhaps stumbled across) this approach to social research in 2010 when a well-known publicist in Britain made headlines by claiming football fans were homophobic and the sport was stuck in the dark ages. This was one of those largely unquestioned statements that some people can make, though without being challenged and asked to produce corroborating evidence. There was none, anyway.

This was one of those areas that seem almost impossible to research with any reliability. The prohibitions are obvious: football has a reputation of having a macho culture in which a particular type of masculinity is thought to predominate. Approaching football fans in or around stadiums, or in pubs, bars and other habitats would have been possible, but probably not productive. Homosexuality may not be a strictly taboo subject, but it is probably not something fans would ordinarily discuss. But this didn't mean they were not interested, nor ready to discuss it, especially after a high-profile figure had besmirched the culture of which they are part.

A questionnaire was launched online. Unlike the more familiar online questionnaires, this contained questions that invited written answers. It included tick boxes, but in conjunction with text boxes that invited open-ended responses. As a way of enriching the experience for respondents, we embedded in the questions links to pertinent newspaper, magazine or journal articles, so that participants could engage more fully with the topics. Respondents could complete the questionnaire on a device, in an environment and at a time of their own choosing. They could, if they wished, disclose their name and email address, but otherwise their identities were protected. There were over 2000 responses.

Critics of self-selecting samples i.e. when participants volunteer to participate, typically point out that the samples may reflect some sort of bias among those who choose to get involved in the research. They also argue that respondents can lie. Of course, the skill of designing a questionnaire lies in catching and sustaining the interest of the potential respondents without loading questions in a way that could result in bias. As for lying: respondents are less likely to answer dishonestly when they're not facing a researcher. Again, the skill of the researcher is paramount: the designers of the questionnaire must remove the motivation to lie.

Contrary to popular assumptions, the project indicated that an overwhelming majority of football fans rejected accusations of homophobia, resented being smeared and argued a players' sexual proclivities were of no interest, but they welcomed the time when a gay player felt emboldened to come out. It was a surprising and, in many ways, counterintuitive finding and one that encouraged us to pursue online projects. We have covered several different subjects, including gambling, racism and violence; though *Screen Society* was the most ambitious project to date: it's aim was to investigate the experience and consequences of what many critics considered our unhealthy fixation with the media.

With a brief like this, we needed a sample limited by only one feature: participants had to engage with screens. Some people who watch television, listen to radio and read newspapers have no interest in any other kind of media. We were not especially interested in them: the people who used their screens habitually were our focus. And, of course, the perfect way to reach them was through their screens.

We divided the research into phases, the first phase being wide-ranging and intended to elicit responses that would reveal the major topics, as users (as opposed to politicians and the media) see them (though, of course, there was an overlap). As with our previous online surveys, we embedded hyperlinks to news stories or academic research papers in questions; these were intended to draw in the participants and challenge them to think in ways that traditional surveys can't. The approach being to upload a questionnaire framed in a way that motivated users to think and note their thoughts on their computers, smartphones, or tablets in any environment of their own choosing, whenever they wished. Their identities were protected and the usual ethical protocols prescribed by the British Sociological Association were followed.

The first phase questionnaire was as free of preconceptions as we could manage: the aim was to explore the issues as Screenagers themselves saw them. So, we used a combination of tick-boxes and open-ended questions that eventually brought us 1500 responses. As in previous projects, we read and interpreted the responses and synthesized several themes expressed by the participants. We then switched to a second phase of the research, in which we asked the participants, again online, to think about and respond to some of the arguments that dominated the first tranche of responses. Just over 350 participants completed this second questionnaire.

As well as designing questions that were based on the responses received at phase 1, we also introduced one new question that was influenced by an article by Sarah Knapton, published in the *Telegraph*. The article claimed that fitness apps (mainly referring to digital Smart Watches that count steps taken and calories burned) could do more harm than good. As this was something that we hadn't considered, we decided to put this argument to our participants. But as well as satisfying their views on this specific issue, we noticed that on occasion, they would elaborate beyond the scope of the question. In doing so they drew our attention to a wider issue that we couldn't ignore—the future of digital health. Participants had done what all good participants do in the end. They, as experts in the field, altered the direction of the enquiry by awakening researchers to the relevance of digital health in the lives of Screenagers.

So, with our attention redirected, we designed a phase two spin-off that asked direct questions about attitudes towards health, exercise and wellbeing, relationships with medical professionals, and projections for the future of healthcare. The responses, as we now came to expect, were detailed and totaled 138.

The third and final phase of the project's empirical part comprised only one question and was underpinned by a slightly different rationale. While the previous phases of the research had set out to canvas views relating to specific topics, the third phase encouraged participants to provide a holistic account of the impact of screens on society and them as individuals. We asked:

How do you think screens have shaped our society, our lives and you?

This was as demanding as the participants wanted it to be; some wrote one-sentence responses, while others postulated at length.

The first key to participant recruitment was project awareness. Hours of engagement in social media platforms, guesting on live streamed debates, articles in print media, and radio interviews raised the profile of the research. Though perhaps most crucially, at each stage of the research we attempted to provide a sense of project ownership to those people that this book is about: Screenagers.

The second key was relevance: subject matter closely connected to what people do or appropriate to what they consider important *now* is bound to get a more fulsome response than subject matter that does neither. Screens are part of everyone's lives, even those who despise them and the world that's been created around them. This is undeniable. Researching a topic like ours is, in a sense, easy.

But that's not to say that everything went smoothly. For example, in phase three, we suspected that we had erred in soliciting responses in online forums that were predominantly populated with males. We first encountered this kind of sampling error in phase one and explored the reasons for an overrepresentation of males among participants. It's worth detouring briefly to explore why so few women responded to the initial questionnaire in phase one.

Women: "They're Not Listened to"

It could be that the response ratio reflects a healthy and understandable scepticism among women, who tend to be targeted more by trolls and other cyber wrongdoers. One woman took the precaution of checking out the researchers beforehand. She explained that women are naturally more cautious of strangers. Another believed there was an additional factor: "There's more than simply the reluctance of women to 'talk to strangers.' I think women are more reluctant to offer their opinions than men, period. It is because our opinions are so routinely disregarded as unimportant in our male dominated societies."

Another female participant added, "sometimes women feel that that they're not listened to, so don't bother with surveys." Social research has its heart in the right place, but rarely puts its shoulder to the wheel when pressing for the changes it proposes.

This was a cogent explanation and one we accepted—until the responses from phase three started to arrive and showed no comparable gender bias. Could we have been responsible for the skew at the start of the project? After all, much of our previous research had sporting themes and many of the forums we previously visited were sports-oriented. We made efforts to balance this by taking a presence in forums in which women were more prevalent. But 68.3% of the responses were from men (or, we should say, people who identified as men). This compared to nearly 75% in Phase 2 and 31% in phase three.

Our tentative conclusion is that women are sensibly more hesitant about replying to online requests and their hesitation is based on the reasons offered by the two participants: caution and scepticism. But when presented with the unusual challenge of a questionnaire featuring a single open-ended question that allowed full expression, they accepted it and responded thoughtfully.

As They See Them

No research method is flawless: self-selecting samples are fraught with problems, particularly online. As the name implies, the sample is decided by the agreement of participants to engage with the project: if they choose to participate, they do so; so there is no way of stratifying the sample, whether by age, gender or any other demographic, unless, of course, it's done retrospectively. The self-selecting sample is almost certain to be unrepresentative. In this instance, only users of screens were able to participate. We intended that.

Contrary to our expectation that most of the sample would be drawn from people under 30, there was an even distribution across the age groups. One of the pitfalls of offering participants to complete questionnaires at their convenience is that a percentage takes the opportunity to exhibit their comic talents. There is probably five percent of participants who respond to questions tongue-in-cheek. But in phase three, about 60% of the responses were written either thoughtlessly or in jest. We excluded these somewhat frivolous responses. Our executive decision was to disqualify these responses, retain the responses that reflected thought and consideration and continue.

Online surveys force researchers to surrender the kind of control they ordinarily have when conducting research face-to-face. If respondents wish to poke fun at the research or address the questions foolishly, they can do so without interference. This is a serious pitfall in tick-box surveys, though less so when questions demand written answers; then the true purpose of the respondent becomes clear.

Remember, the final phase of the project was a single question, inviting reflection and thoughtful responses. The fact that so many of the responses were reflective and often studious suggested that many took the opportunity to explain their thoughts.

We were able to determine how many people looked at the question in total and compare this with the number who decided to participate: the conversion rate was just ten percent, meaning that nine out of ten viewers were put off by the prospect of sharing their thoughts through text as opposed to tick-boxes. The final number of participants prepared to commit their thoughts online to phase three was 53. So, even allowing for the jokers, this provided us with a varied and multi-layered bank of data—on which we've drawn throughout this book, but especially in Chapter 13.

As we pointed out earlier, no methodology is perfect and ours is no exception. It is, however, as close as we could design to an instrument that avoids imposing ideas, theories or arguments and simply allows participants to present their views, describe the ways in which they feel their lives are shaped by screens and express the relevant issues in their lives as they—rather than researchers—see them.

Every decent research project goes in unexpected directions. If it goes according to plan, then it's probably a bit too pasteurized—safe for consumption but hardly thrilling. But the weaknesses of being led astray are obvious: research can lose focus, samples can become unrepresentative and data can be adulterated. All the same, we think it's been a risk worth taking. We set out to discover how screens were shaping our society. Who better to ask than the people who constitute that society?

References

Cashmore, E., & Cleland, J. (2011). Glasswing butterflies: Gay professional football players and their culture. *Journal of Sport and Social Issues, 35*(4), 420–436.

Cashmore, E., & Cleland, J. (2012). Fans, homophobia and masculinities in association football: Evidence of a more inclusive environment. *British Journal of Sociology, 63*(2), 370–387.

Knapton, S. (2017, February 21). The 10,000 steps a day myth: How fitness apps can do more harm than good. *Telegraph*. Available at: bit.ly/2haHhAL. Accessed November 2017.

Index

© The Editor(s) (if applicable) and The Author(s) 2018
E. Cashmore et al., *Screen Society*, https://doi.org/10.1007/978-3-319-68164-1